Deflating Mental Representation

The Jean Nicod Lectures

Francois Recanati, editor

The Elm and the Expert: Mentalese and Its Semantics, Jerry A. Fodor (1994)

Naturalizing the Mind, Fred Dretske (1995)

Strong Feelings: Emotion, Addiction, and Human Behavior, Jon Elster (1999)

Knowledge, Possibility, and Conscsiousness, John Perry (2001)

Rationality in Action, John R. Searle (2001)

Varieties of Meaning, Ruth Garrett Millikan (2004)

Sweet Dreams: Philosophical Obstacles to a Science of Consciousness, Daniel C. Dennett (2005)

Things and Places: How the Mind Connects with the World, Zenon W. Pylyshyn (2007)

Reliable Reasoning: Induction and Statistical Learning Theory, Gilbert Harman and Sanjeev Kulkarni (2007)

Origins of Human Communication, Michael Tomasello (2008)

The Evolved Apprentice: The Nicod Lectures 2008, Kim Sterelny (2011)

Deflating Mental Representation, Frances Egan (2025)

Deflating Mental Representation

Frances Egan

The MIT Press
Cambridge, Massachusetts
London, England

The MIT Press
Massachusetts Institute of Technology
77 Massachusetts Avenue, Cambridge, MA 02139
mitpress.mit.edu© 2025 Massachusetts Institute of Technology

The MIT Press would like to thank the anonymous peer reviewers who provided comments on drafts of this book. The generous work of academic experts is essential for establishing the authority and quality of our publications. We acknowledge with gratitude the contributions of these otherwise uncredited readers.

This book was set in Stone Serif and Stone Sans by Westchester Publishing Services. Printed and bound in the United States of America.

Library of Congress Cataloging-in-Publication Data

Names: Egan, Frances, author.
Title: Deflating mental representation / Frances Egan.
Description: Cambridge, Massachusetts : The MIT Press, [2025] | Series: The Jean Nicod lectures | Includes bibliographical references and index.
Identifiers: LCCN 2024022584 (print) | LCCN 2024022585 (ebook) | ISBN 9780262551601 (paperback) | ISBN 9780262381963 (epub) | ISBN 9780262381970 (pdf)
Subjects: LCSH: Mental representation.
Classification: LCC BF316.6 .E43 2025 (print) | LCC BF316.6 (ebook) | DDC 153.2—dc23/eng/20240804
LC record available at https://lccn.loc.gov/2024022584
LC ebook record available at https://lccn.loc.gov/2024022585

10 9 8 7 6 5 4 3 2

EU product safety and compliance information contact is: mitp-eu-gpsr@mit.edu

Contents

Acknowledgments

I am grateful to the Centre National de la Recherche Scientifique (CNRS) and the Institut Jean Nicod for inviting me to give these Jean Nicod lectures and for hosting me for a very pleasant and productive few weeks in Paris in November 2021. Thanks to Frederique de Vignemont, Francois Recanati, Roberto Casati, and audiences at the lectures for profitable and spirited discussion. Special thanks to Pierre Jacob for making abundant time for stimulating conversation (much of it at the Café Le Rostand) during my visit and for valuable follow-up correspondence.

I am indebted to Oron Shagrir and Peter Slezak for helpful comments and suggestions on the penultimate version of the manuscript and to Ken Aizawa, Alex Grzankowski, Chris Hill, Adam Pautz, Mark Sprevak, and Neil Van Leeuwen for very useful feedback on individual chapters. Thanks also to four anonymous referees for their careful reading of the manuscript.

I have benefited from interactions with audiences at various conferences and colloquia where I presented versions of this material. I have discussed most of the material, in one form or another, with Albert Newen and colleagues during regular visits to the Center for Mind and Cognition at Ruhr University/Bochum. Thanks also to participants in the graduate seminar on Intentionality that I co-taught with Brian McLaughlin at Rutgers University in fall 2022.

My greatest debt is to my spouse, Bob Matthews, for incisive comments and suggestions, and for his constant love and support.

Series Foreword

The Jean Nicod Lectures are delivered annually in Paris by a leading philosopher of mind or philosophically oriented cognitive scientist. The 1993 inaugural lectures marked the centenary of the birth of the French philosopher and logician Jean Nicod (1893–1931). The lectures are sponsored by the Centre National de la Recherche Scientifique (CNRS), in cooperation with the Ecole des Hautes Etudes en Sciences Sociales (EHESS) and the Ecole Normale Superieure (ENS). The series hosts the texts of the lectures or the monographs they inspire.

Introduction

The topic of these lectures is *mental representation*, the capacity of some mental states to represent objects, properties, and states of affairs outside themselves. For example, sitting at my desk right now in New Jersey I can grasp, visually, the coffee cup in front of me, but I can also think about Paris, about Napoleon, about the galaxy Alpha Centauri, about the property of justice, about the number 2. An obvious question is how mental states get their content. What is it about states of me—of my mind or brain—that enables them to represent not only objects in front of me but also objects remote in space and time, and abstract objects outside of space and time altogether, things that cannot causally affect me here and now as I am thinking about them? This is *the problem of intentionality*.

Contemporary philosophical accounts of mental representation answer the question by positing a substantive relation—what we might call the *representation relation*—holding between a mental state and what it is about, and then attempting to characterize the relation in such a way that it picks out just the represented object or property, and not lots of other things that may be in the vicinity. I recommend a different approach, one that requires appreciating the role that mental representation plays in the various domains in which it figures—as an explanatory posit in the sciences of the mind, in our commonsense practices of predicting and explaining each other's behavior, and in our thought and talk about perceptual experience. In the following chapters, I examine the role that appeal to mental representation plays in each of these domains.

I propose what I call a *deflationary* account of mental representation, characterized by the following three claims:

1. Construing a mental state as a *representation* does not presuppose a special, substantive relation—what we can call a *representation relation*—holding between the mental state and what the representation is about, for example, the nearby coffee cup, or Paris or Napoleon.
2. Content is not an essential property of mental states—that is, the very same kind of mental state might have had a different content, or no content at all.
3. The attribution of content to mental states is always pragmatically motivated. It serves to characterize, for certain uses and purposes, features of the mind that are not themselves intentional. Most generally, the attribution of content to mental states allows us to *model* mental states and mental processes—which are not publicly observable or accessible—in ways that we already understand well. As I put it, content serves to *gloss* these features for the intended purposes.

In the following chapters I characterize *what it is* that is glossed by content in the various domains in which it is attributed and explain the function(s) that the gloss is serving.

The first two chapters focus on the scientific investigation of the mind. The so-called cognitive revolution of the last sixty years has seen the fruitful application of methods from mathematics and computer science to the study of the mind. Computer models of vision, language understanding and production, deductive reasoning, and other cognitive capacities promise to shed light on how humans interact successfully with each other and with their environment. In articulating their models, computational psychologists and neuroscientists often speak of *representation*. In chapter 1, I develop and defend a deflationary proposal about how such talk is to be understood. The upshot is that content attribution in these sciences is irredeemably pragmatic.

In arguing that content attribution is pragmatically motivated, dependent in part on the interests and explanatory goals of theorists, I reject the attempt by philosophers of mind to *naturalize* intentionality by specifying, in non-semantic and non-intentional terms, sufficient conditions for a mental state to have its particular content. In chapter 2, I explain how to understand the project of naturalizing the mind on a deflationary construal of mental representation. I argue that only by eschewing essential commitment to such notions as *meaning* and *intentionality* can computational neuroscience and computational psychology hope to serve as a naturalistic foundation for the study of mind. But these sciences cannot simply throw such notions away: they are needed to demonstrate that the mathematical and mechanical accounts of cognitive capacities furnished by the sciences are indeed explanations of genuinely mental phenomena—to demonstrate that the target of the naturalistic inquiry is indeed *the mind*.

In the last two chapters, I argue that beliefs (chapter 3) and perceptual experiences (chapter 4) are best understood not as states of subjects bearing a substantive relation to their worldly objects, as the most popular philosophical accounts in these domains claim. They are rather to be understood as properties of subjects that are modeled by aspects of external reality. Beliefs, in particular, are modeled as linguistic objects that have syntax as well as content and truth conditions. In attributing content and truth conditions to beliefs, we *gloss* them (and the other attitudes) as linguistic objects with such properties, but without commitment to the idea that they literally have these properties. Rather, they have properties that have linguistic properties as images. This scheme for modeling beliefs in linguistic terms enables the prediction, explanation, and regulation of behavior in our quotidian practices.

In the philosophy of perception, *representationalism* is the view that all phenomenological differences among perceptual experiences—differences in "what it is like" for the subject to undergo the experience—are representational differences: that is, differences in content. In the final chapter, I defend an alternative view that I call *external sortalism*, inspired by traditional adverbialism about perceptual experience, and according to which experiences are not essentially representational, though they can be attributed content that serves various important pragmatic purposes. The central idea of external sortalism is that the external world serves as a model for sorting, conceptualizing, and reasoning surrogatively about perceptual experience. On this view, contents are again construed as a kind of *gloss*. The view allows us to retain what is attractive about representationalism—namely, that perceptual experiences can be evaluated for accuracy, and hence for whether they can serve as a rational basis for belief, but without problematic commitment to the idea that they bear a substantive, representational relation to external objects and properties and that this relation determines their phenomenal character.

Taken together, the chapters present a coherent picture of the attribution of mental content in computational cognitive science, in our ordinary predictive and explanatory practices, and in our understanding of perceptual experience. The deflationary account is not intended to be revisionist about the role of representation in either science or ordinary life, though I challenge the standard *philosophical* interpretations of both. I am not holding either science or ordinary practice to some philosophical, *a priori* standard about what counts as a legitimate entity or property, or a genuine explanation. These practices are surveyable, and the deflationary view is intended to be empirically adequate to these practices.

1 Representation in Cognitive Science

Cognitive science is replete with talk of *representation*. Structures posited in computational theories of vision are said to represent edges in the world. Neurons are said to represent elements of their receptive fields. Despite the widespread use of representational talk, there is surprisingly little consensus about how such claims are to be understood. Should representational talk be taken literally, or is it just a useful fiction? My goal in this chapter is to explicate the role that appeal to representation plays in computational neuroscience and computational cognitive psychology, the sciences that aim to explain our cognitive capacities. I argue that neither a robustly realist nor a fictionalist interpretation of representational talk is appropriate. I call the right interpretation *deflationary*. It couples a realist construal of the structures and processes posited by computational theories with a pragmatic account of representational content attributed to them. Most importantly, the deflationary view denies that a substantive representation relation holds between the structures posited in the theory and the objects and properties they are taken to represent.

1.1 Preliminaries

Before we turn to *mental* representation, let's get a fix on the topic by considering what it is for something to count as a representation more generally. The central examples of representations are publicly observable, such things as spoken utterances, written text, maps, photographs, drawings, graphs, pie charts, and so on. Characteristic of all these examples is a distinction between the *vehicle* of representation—that is, a physically realized entity of some sort that has causal powers, and its *content*, in other words, the meaning, message, or information conveyed. A representation must have a vehicle; meanings are not free-floating but are anchored to or carried by structures or entities of some sort. The meaning of an utterance is carried by an acoustical sequence, the meaning of written text by sequences of graphic symbols. Equally important is that having content entails having *correctness conditions*, in other words, conditions in which the representation counts as correct or accurate, and so, conversely, conditions in which the representation would be inaccurate, *misrepresenting* the way things are. If Jack is standing on a box, then a photograph may misrepresent him as being taller than Jill. A map of France will misrepresent Marseilles as north of Paris if the labels have been switched. To count as a genuine representation, a mental state or entity should satisfy two conditions: (1) it should support a distinction between vehicle and content and (2) it should be capable of misrepresentation. If computational cognitive models do not posit something like the central cases—with these characteristic features—then there is no good reason to say that they posit representations. Of course, there will be differences; for example, mental representations are in the head, not publicly observable. And while public representations get their meaning in part by *convention*—fixed by the intentions of users—mental representations must acquire their meaning in some other way. Any theory that posits mental representations presumes that a plausible story can be told about how they get their meaning.

We can turn now to our question, considering whether computational cognitive models posit entities sufficiently like central cases of representations in relevant respects. Here are two popular answers:

1. Robust realism—representational talk is to be taken literally: mental representations are as real as written texts, maps, and so on.
2. Fictionalism—representational talk is just a useful fiction, serving various purposes, primarily the prediction and systemization of behavior, but is not to be taken literally.

I will argue that both views are mistaken, though there is an element of truth in each. Very roughly, robust realism is right about vehicles and wrong about content, and fictionalism is wrong about vehicles, and "on the right track" about content. I will argue for a third possibility, a view I call *deflationism*, according to which mental representations are real states or structures that are given what I call an *intentional gloss*. Before turning to the status of representational talk, let us consider how we should think about computational models.

1.2 Understanding Computational Models of Cognitive Capacities

One motivation for fictionalism is the widely held view that computational descriptions applied to human cognition are mere metaphors. Minds, it is sometimes claimed, aren't really computers. Barrett (2011) takes this to be obvious:

> Some people go so far as to argue that the (human) brain is not just analogous to a computer in a strictly metaphorical sense, but that it actually is a computer in the sense that it takes an input, processes it in various ways, and then produces a specific output. . . . Our use of the computer metaphor is so familiar and comfortable that we sometimes forget that we are dealing only with a metaphor and that there may be other, equally interesting (and perhaps more appropriate) ways to think about brains and nervous systems and what they do. (Barrett 2011, 114)

Similarly impressed with the apparently yawning gap between computers and brains, Bennett and Hacker, in their *Philosophical Foundations of Neuroscience* (2003), say the following about computers:

> We use computers to produce the results of a calculation—just as we used to use a slide-rule or cylindrical mechanical calculator. Those results are produced without anyone or anything literally calculating—as is evident in the case of a slide-rule or mechanical calculator. In order literally to calculate, one must have a grasp of a wide range of concepts, follow a multitude of rules that one must know, and understand a variety of operations. Computers do not and cannot. (2003, 7–8)

Bennett and Hacker claim that calculating requires *grasping* concepts, *knowing* rules, and *understanding,* all of which, they claim, are beyond the capacities of computers. Given that a common dictionary definition of "calculate" is *to compute,* and conversely, that "compute" is commonly defined as *to calculate,* Bennett and Hacker's claim has the interesting consequence that computers do not literally compute. But rather than appealing to intuitions about the ordinary meaning of "compute," let's consider some implications of the view that computational descriptions applied to human cognition are mere metaphors.

Of course, a computational description *might* be a metaphor of sorts, not to be construed as a literal description of the brain.

A realist interpretation of a computational model is not always appropriate. For example, some Bayesian theorists claim that their models are intended simply to systematize and predict behavior, not to describe how the brain actually works.[1] But if the purpose of the theory is to *explain* a cognitive capacity—that is, to explain how the brain accomplishes some cognitive task—then a metaphor will not do.[2] The metaphor view is mistaken as an account of computational models that are proposed as genuinely explanatory accounts of cognition, as many are. A computational model that characterizes a neural system as computing integration, say, purports to literally describe neural processes. According to such accounts, the brain computes the specified function in the same sense as a hand calculator or an iPad does.

To be sure, a computational characterization involves *idealization*, but a computational description of a neural system involves no more idealization than does a computational description of any other physical system. A calculator's actual capacity is limited by the size of its display. Biological systems have memory and other limitations. Despite idealization, the computational characterization of a system as computing, say, *addition* is intended to be literally true of both the brain and the calculator. The explanatory credentials of the model depend on this. If it is just a metaphor, there is no explanation.

Let me issue a couple of caveats. At a very early stage of theorizing, the model may not be much more than a promising guess, with very little empirical support. Having proposed that a neural system computes a particular function, the theorist may have little idea how the computation is realized in the brain. At this stage, the model may be the merest of "mechanism sketches," in the sense explicated by Piccinini and Craver (2011). Still, assuming that the goal is to explain a capacity of the system, it is a hypothesis about what the brain is actually doing, and so a literal interpretation is appropriate.[3]

Secondly, there is always a question about which aspects of an empirical theory should be interpreted realistically, and which aspects should be construed simply as features of the representational scheme used to articulate the theory. An analogy illustrates the general point. Map projections allow us to represent the curved surface of the earth on a plane. Different projections preserve different properties of the earth's surface—area, distance, shape, direction, and so on—but no single projection preserves all metric properties. *Mercator projection* is the standard used by mariners because, locally, it accurately represents the angle or bearing between two locations (the rhumb line), although it distorts size and shape increasingly as one moves away from the equator. Greenland looks as big as North America. It would be a mistake to conclude that size on a Mercator projection represents actual size. Colors on a map may represent topographical features or political boundaries; alternatively, color may have a nonrepresentational function, perhaps just making the map easier to read. The general lesson is that, as with maps, one needs to be careful in attempting to read the empirical commitments of a theory from its explicit formulation. This point will turn out to be particularly significant when considering the commitments of a computational theory to representation.

We will now turn to the question: *What is it for a physical system (brain or computer) to compute a function?* Let's consider a physical system that computes addition, depicted schematically in figure 1.1.[4]

Most generally, a physical system has the capacity to compute the addition function just in case there exists a mapping from physical states of the system to numbers, such that

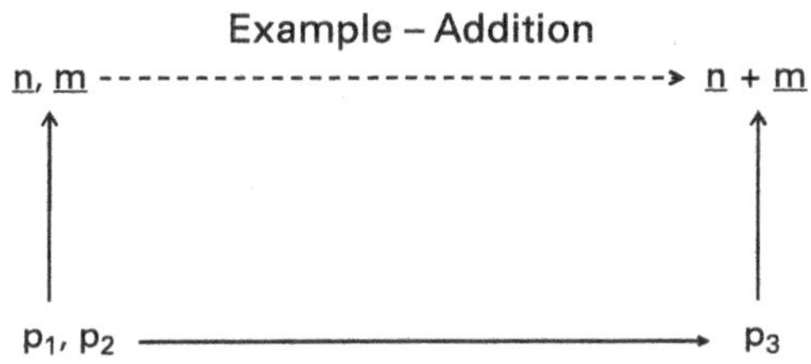

Figure 1.1
Example—an adder.

physical states related causally (p1, p2, p3 in the bottom span of figure 1.1) are mapped to numbers n, m, and n+m related as addends and sums (the top span). The physical states may be states of brains, states of silicon-based systems, or some other kind of physical states.

The bottom horizontal arrow of figure 1.1 depicts a *causal* relation; the top horizontal arrow depicts a mapping of arguments of the function into values. Whenever the system is in the physical states specified under the mapping as the numbers n and m, it is caused to go into the physical state specified under the mapping as n + m. For example, if the system is in the physical state specified as 3 and the physical state specified as 2, then it will go into the physical state specified as 5, and similarly for other addends and sums.

The mapping depicted in figure 1.1 specifies the causal organization underlying and supporting the physical system's capacity to add. Computational models characterize this causal organization in a compact and perspicuous way by specifying the function (in the mathematical sense) that the system computes (the top span), and then, in a fully articulated model of a human cognitive capacity, specifying the neural states that realize the arguments and values of the function (the bottom span). Importantly, no physical system just adds; every physical system has lots of causal organization, for example, some of it is responsible for maintaining the system's physical integrity over time. The mapping in figure 1.1 singles out the causal organization specifically responsible for the system's capacity to add.

Returning to the topic of representation, we could construe physical states that are mapped to numbers to be *representations* of those numbers, construing these physical states to be representational vehicles, but we can give a more perspicuous rendering of the system's representational capacities by complicating the picture (figure 1.2).

If a model posits representations, then there will be two mappings. The bottom mapping (f_R) specifies the *vehicles* of representation, which in our addition example are numerals but generally will be structures of some sort (hence S_i). This bottom mapping specifies how vehicles are physically realized, and so I will call it the *realization function*. The top mapping (f_I) specifies the *contents* carried by the vehicles—what is represented—which in our example are numbers. I will

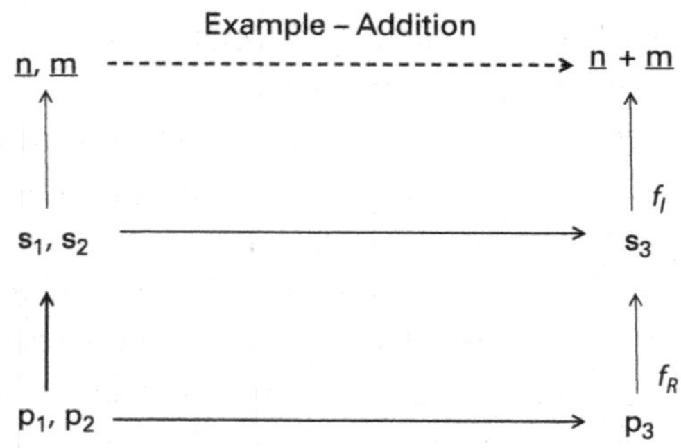

Figure 1.2
An adder—complicating the picture.

call the top mapping the *interpretation function*. The bottom two horizontal arrows depict causal relations, the middle at a higher level of abstraction, abstracting away from much of the physical detail at the bottom level. As before, the top horizontal arrow depicts a mapping of arguments of the function into values. So, to repeat, the *realization function* (f_R) specifies the vehicles of representation—that is, which physical states count as which structures. The *interpretation function* (f_I) specifies their content.

Before proceeding, I will make a point about vehicles and a point about contents. We should be liberal about what counts as a vehicle. Vehicles may be discrete symbols, as in a language-like system, but they could also be, say, characteristic patterns of activation of a neural network. The realization function (f_R) will abstract away from some properties of the realizing physical states and group them together by their role in cognitive processing. Typically, I will use "structures" as a blanket term for the sorts of things that function as vehicles. "State" might be more accurate—it is more inclusive, more obviously including such things as characteristic patterns of activation of a network—but "structure" serves to keep us in mind of vehicles.

A point about contents: A computational theory construes the vehicles specified by f_R as representing the arguments and values of the computed function. In our addition example, these vehicles represent *addends* and *sums*. I call the content determined by the specified function *mathematical content*. In virtue of characterizing the system as computing the function the theory is committed to such content attributions. So, a computational theory that posits the causal organization depicted in figure 1.2 in an explanation of a physical system's capacity to add posits real physical structures that represent addends and sums.

The explanatory strategy explicated here is pervasive in computational psychology and computational neuroscience. I will briefly describe a few applications drawn from a variety of cognitive domains:

1. Marr's (1982) theory of early vision purports to explain edge detection by positing the computation of the Laplacian of a Gaussian of the retinal array. In effect, the mechanism computes a smoothing function to facilitate the identification of edges in the scene and to eliminate noise. It takes as input intensity values at points in the image and calculates the rate of intensity change over the image. Its inputs represent intensity values, and its outputs represent the rate of intensity change.
2. Ullman (1979) hypothesizes that the visual system recovers the 3D structure of moving objects by computing a function from three distinct views of four non-coplanar points on the object to the unique rigid configuration consistent with the points. Inputs to the process represent points in space and outputs the unique rigid structure connecting the points. In a world such as ours, where objects are typically rigid in translation, the process enables the recovery of 3D structure from minimal information.
3. Shadmehr and Wise's (2005) computational account of motor control purports to explain how a subject is able to grasp an object in view. They decompose the overall task into three distinct computations. First, a network in the posterior parietal cortex computes *forward kinematics*—that is, hand location, in visual coordinates, from proprioceptive information from the arm, neck, and eye muscles, and information about the angles of shoulder and elbow joints. Next, a network in the premotor cortex computes

a movement plan by computing the *difference vector*—that is, the displacement of the hand from its current location to the target location. Finally, a network in the primary motor cortex computes *inverse kinematics,* the computation from the high-level movement plan, in eye-centered coordinates, to a required change of joint angles. Reaching and pointing movements involve continuous monitoring of target and hand location, with the goal of reducing the difference vector to zero.

4. In a well-known example from animal cognition, Gallistel (1990) purports to explain the Tunisian desert ant's impressive navigational abilities by appealing to a mechanism for *path integration.* The mechanism continuously computes the displacement vector to the nest from any point along the ant's foraging trajectory, integrating information about direction and speed. The process enables the ant, after completing its search, to turn and head directly back to the nest.
5. Seung (1996; 1998) and Seung et al. (2000) hypothesize that the brain keeps track of eye movements during saccades by deploying an internal integrator. They posit a multi-stable recurrent network located in the brainstem that takes as input transient eye velocities and gives as output persistent eye positions. It does so by accumulating input pulses, adding or subtracting (depending on the direction of movement) new inputs from the previous summation. Inputs to the mechanism represent eye velocities, and outputs represent stable eye positions.

These mechanisms all compute well-defined mathematical functions that play crucial roles in cognitive processes, and their input and output states represent the arguments and values of the functions, respectively.[5] In attributing a computational capacity or competence to a physical system—to compute vector displacement, or a smoothing function, for example—a computational theory, in particular, the mapping from physical states of the system to the arguments and values of the computed function, supports attributions of correctness and mistake. Just as the normal functioning of the system—correctly computing the specified mathematical function—explains the subject's success at a cognitive task in its normal environment, so too a malfunction would explain its occasional failure. Consider again the addition example: If the adder were to go into the physical state interpreted under the mapping as 99—when in prior states interpreted as 53 and 47—it would miscompute or make a mistake. We can say that it *misrepresents* the sum of 53 and 47 as 99. On the other hand, if its battery died or it was dropped in the bathtub it would most likely go into a physical state not interpreted under the mapping at all; it would not miscompute, it would *malfunction.*[6] The specification of the function computed (here, addition) and the mapping of physical states to elements of the computed function (here, to addends and sums) partitions the behavioral space into (1) correct computations, (2) miscalculations or mistakes, and (3) malfunctions (i.e., states that receive no interpretation under the mapping). The specification of the function computed and the interpretation together provide the norm necessary to underwrite this three-fold distinction. And so, given that both natural and artifactual adders can be characterized computationally, an appeal to *biological function* is not needed to underwrite the distinction between well-functioning and malfunctioning in a natural system, contra Neander (2017) and Garson (2019). The addition example illustrates how a distinction between well-functioning and malfunctioning arises in computational cognitive science. It also illustrates how a computationally characterized mechanism is capable of misrepresentation.

1.3 The Attribution of *Cognitive* Content

I will have more to say about what I am calling *mathematical content* in chapter 2, but the debate about how to understand representational talk is not about mathematical content. When computational theorists use representational talk, they are typically not referring to the arguments and values of the functions computed by the mechanisms they posit. Rather, theorists talk of the visual system representing edges in the world, and of "place cells" in rat brains representing locations in the rat's local environment.[7] In other words, they talk of internal states representing external, distal objects and properties. I call this external, distal content *cognitive content* (as opposed to *mathematical content*) because it has a more obvious connection to the cognitive capacities that are the explanatory targets of computational theories, for example, such capacities as the capacity to detect the 3D structure of nearby space, the capacity to navigate one's environment, and so on. In the rest of this chapter, I will focus on this more familiar cognitive content.

1.3.1 Resolving Indeterminacy

The central question here is how cognitive content is determined—that is, how the internal structures posited in computational theories get their external, distal meanings. There has been no shortage of proposals, most aiming to respect a *naturalistic constraint*, specifying the relation between an internal structure and what it is about—the *representation relation*—without reference to semantic or intentional properties. Computational theorists aim to provide a naturalistic explanation of the representational capacities of agents; appeal to semantic or intentional notions in their theories would frustrate this aim, assuming the very phenomenon to be explained. Naturalistic accounts of cognitive content—more generally, naturalistic theories of *intentionality*[8]—typically characterize the representation relation as a kind of *tracking* relation: the internal structure is said to track or covary with some external object or property, and that relation determines its meaning. The most popular naturalistic proposals fall into three broad categories.

According to *information-theoretic* accounts, an internal structure S means *cat* if S is caused by the presence of a cat, and certain further conditions obtain.[9] Further conditions are required to allow for the possibility of misrepresentation—that is, for the possibility of some S-tokenings *not* caused by cats. Further conditions may include the requirement that during a privileged learning period only cats cause S-tokenings (Dretske 1981) or that non-cat-caused S-tokenings depend asymmetrically on cat-caused S-tokenings (Fodor 1987; 1990). A well-known problem for information-theoretic theories is that everything in the causal chain from the presence of a cat in the distal environment to the internal tokening of S, including cat-like patterns in the retinal image—which will differ from case to case—may appear to satisfy the condition, and so fall into S's extension, but the internal structure S does not represent all of these things.

Teleological accounts hold that an internal structure S means *cat* if it has the biological function of indicating cats and was selected by evolution because it served this function. The view was first developed and defended in Millikan (1984), and there are now many interesting variations on the central idea.[10] I will have more to say about prominent teleosemantic accounts later in the chapter.

A third type of tracking theory appeals to *structural similarity* or homomorphism—the type of relation that holds between a map and the objects and properties it represents. Cummins (1989), O'Brien and Opie (2004), Ramsey (2007), and Shagrir (2012; 2022) among others have proposed variations on this idea. Since similarity is a symmetric relation but the representation relation is not, any account that attempts to ground content in similarity will need supplementation by appeal to something like *use*. Items in the map represent items in the world and not the other way around because the map is *used* by the system to navigate the world.

Naturalistic accounts of cognitive content notoriously have trouble specifying determinate content for internal representations, a condition that, we shall see, is necessary for a structure to count as a genuine representation. The various tracking relations that have been proposed all fail to pick out a unique external condition, instead supporting the attribution of multiple different contents. The problem is most clearly illustrated by reference to Lettvin et al.'s (1959) frog/fly example. They famously asked, "What does the frog's eye tell the frog's brain?" When a fly crosses the frog's visual field, it engages a mechanism that causes the frog's tongue to lash out, frequently catching the fly. A structure in the frog's brain *represents* the external condition, but what, precisely, is its content? There are a number of candidates, all of which have been proposed by teleosemantic theorists, and all of which have a plausible teleological rationale:

a. The structure means *bug*, because it is typically bugs, in the frog's normal environment, that trigger the mechanism, and so its function is to enable the frog to catch bugs (Millikan 1989).

b. The structure means *frog food,* because the biological function of the process is to procure food for the frog. If it hadn't played an important role in procuring food, it wouldn't have been selected for by evolution (Sterelny 1990, Millikan 1991).

c. The structure means *small, dark moving thing,* because by detecting small, dark moving things in the frog's normal environment, the mechanism thereby catches bugs and thereby procures food for the frog.[11,12]

There is a naturalistic rationale for each of these competing contents—the internal structure tracks each of these things. But appeal to a naturalistic tracking relation doesn't pick out a unique content.

One may be tempted to react here with a shrug. Why does indeterminacy matter? Why not simply say that the internal structure represents *all* of these things? There is, of course, nothing wrong with saying that the mechanism *detects* bugs, frog food, small dark moving things, and maybe lots of other things as well. Detection, unlike representation, is not an intentional relation; it is extensional and factive. But indeterminacy does matter if the structure's firing is to be construed as representing. Recall that a characteristic feature of a representation is that it can misrepresent, but if the structure doesn't have a determinate content and so determinate correctness conditions, then it is impossible to draw a line—even a blurry line—between a correct tokening of the structure and a mistake. Consider again the three candidate contents. Each proposal draws the line in a different place. Suppose that the structure represents *bug*; then if a food pellet engages the mechanism, then the structure's tokening misrepresents the pellet as a bug. But if it represents *food*, this will be an accurate representation. And if the structure represents *small, dark moving thing*, then it accurately represents even when it catches a BB lobbed at the frog in the laboratory.

Nonetheless, there is something right about the objection that the structure has *all* the proposed contents: any of these contents—*bug, frog food, small dark moving thing*—might plausibly be attributed, just *not all at once*. How is the theorist to decide among them? I maintain that we get determinate content only by appealing to the theorist's explanatory aims and goals. In other words, the justification for selecting one content over the others is *pragmatic*, but if this is right, then determinacy of cognitive content is relative to explanatory aims and goals.

Consider once again the candidate contents. If the theorist's goal is to explain the frog's role in its ecological niche, then she is likely to say that the structure represents *bug*. Doing so would situate the frog's behavior in its larger environmental context, relating it to other organisms that occupy that context. Alternatively, if the goal is to explain how the mechanism contributes to the frog's survival and well-being, and hence why it might have been selected for, then the content *frog food* is more appropriate. It is the bug's role as food for the frog that is important to that project. But if the goal is to explain how the frog's visual mechanism works then the theorist is more likely to say that the structure represents *small, dark moving thing*. The fact that the only small, dark moving things in the frog's normal environment are bugs, or food, isn't relevant to this last project, since the visual mechanism cannot discriminate non-bugs from bugs, or non-food from food, when they are all small, dark moving things.

The general point is that the indeterminacy of content is resolved only by reference to specific explanatory interests—that is, to theorists' goals and intentions—most plausibly, I suggest, to *collective* goals and intentions of scientific communities. Once we have determinate content, we have a way of partitioning the system's responses into correct applications and mistakes (i.e., *mis*representations). But we get determinate content only relative to the pragmatic choices of theorists.

It may be objected that whether the brain genuinely represents or not—whether it is doing something enough like our central cases to count as representation—is an *objective* matter, not something to be determined by reference to the goals and pragmatic choices of theorists.

There is some truth to the objection. Computational theories posit structures that play specified causal roles in cognitive processes. It is a purely objective matter whether there are such structures and whether they behave in the way the theory specifies. If these structures are assigned content, then they function as representational vehicles. But the idea that these structures are assigned determinate content independently of the pragmatic choices of theorists should be resisted. A close look at two prominent naturalistic proposals makes this clear.

1.3.2 Two "Naturalistic" Proposals Compared

Neander (2017) and Shea (2018) are the most detailed and carefully argued recent naturalistic semantic proposals.[13] Both appeal to empirical work in cognitive science and reject appeals to intuition. But, in fact, neither account is naturalistic in the advertised sense—that is, neither specifies non-intentional and non-semantic conditions for the internal structure to have its assigned content. Each account appeals tacitly to pragmatic considerations to privilege its choice of content. I will spell this out briefly.

First, a caveat. Lettvin et al.'s seminal (1959) paper frames the issue in terms of "what the *frog's* eye tells the *frog's* brain," and subsequent discussion has followed suit. Neander (2017) justifies her account by appeal to extensive research on the anuran toad visual system. Toads and frogs are, of course, different in many respects, but for present purposes, the differences do not matter. Both belong to the order *anura*. For ease of exposition and to facilitate comparison of Neander's and Shea's accounts, I will elide Neander's talk of toads and Shea's talk of frogs, referring in the following discussion to *anurans*, which subsumes both toads and frogs.

Let R be the retinal ganglian cells in the anuran brain that respond to a fly crossing the animal's visual field and engage its tongue-lashing mechanism. R is the representational vehicle; the issue here is what is its content. Neander (2017) claims that R's firing carries the content *moving wormlike stimuli (there, now)*. Shea (2018) argues that R's firing means *nutritious flying object (there, now)*. The accuracy conditions for the two attributions are different: a moving wormlike stimulus that is not nutritious for the anuran will count as an accurate representation on Neander's account but as a misrepresentation on Shea's. Is one right and the other wrong about R's content? Let's look more closely at the two accounts.

Neander notes that anura typically orient toward prey-like stimuli, in particular, toward "wormlike" stimuli of constant width and varying length that move parallel to their longest axes. This is the *sign signal,* the configuration of environmental features that triggers the response. The activation of a particular class of cells in the tectum—known as "T5-2" cells—produce directly, without further processing, the appropriate orienting response and are thought by neuro-ethologists to correspond most closely to the recognition of prey-like stimuli.

As Neander puts it, "This is their candidate for the relevant representation, whose content is debated by philosophers" (2017, 108). The activation of T5-2 cells in the optic tectum is R, the representational vehicle. "Moving wormlike stimulus" captures the precise stimulus conditions for the activation of T5-2 cells—that is, for R's firing. *Moving wormlike stimulus,* Neander says, is R's content.

Shea's preferred content attribution—*nutritious flying object*—captures the mechanism's contribution to anuran well-being, what he calls its *task function*. The mechanism's role in getting nutrients into the anuran explains why it would have been selected by evolution. Shea cites what he terms *transparency of explanation* as decisive for determining R's content:

> In the classic example of the . . . [anuran] fly-catching mechanism, the correlation of retinal ganglion cell firing (R) with little black things figures in an explanation of how the system was stabilized by evolution, but that explanation also mentions the fact that being a little black thing (condition C) correlates with being a nutritious flying object (condition C′). Without that background correlation, *it would be opaque how the correlation between R and C enabled . . . [anurans] to achieve an evolutionarily beneficial outcome.* So the role of the R-C correlation in that explanation is mediated. There is another explanation of stabilization that adverts directly to the correlation between R and nutritious flying objects (C′). The role of the R-C′ correlation in that explanation is unmediated. (2018, 84, emphasis added)

According to Shea, the tight ("unmediated") connection between R's firing and the task function for the system—getting nutrients—in the explanation of the mechanism's role in selection justifies attributing the content *nutritious flying object* to R.

But the "unmediated" connection to which Shea refers here is in the theorist's explanation, not in the world. Little black things (C) in the anuran environment *just are*, enough of the time, nutritious flying objects (C′). Moving wormlike stimuli, in that environment, typically *are* nutritious flying objects. The naturalistically specifiable facts provide no reason to favor Shea's content attribution—*nutritious flying object*—over Neander's *moving wormlike stimuli*. Attributing the content *nutritious flying object* makes for transparent as opposed to opaque explanation only if the goal of the theorizing is explaining how the system was stabilized by evolution. We get the opposite verdict if the explanatory target is explaining how the mechanism works in the here and now. In the latter case, construing R as representing *moving wormlike stimulus* makes for more transparent explanation, since that content attribution better captures the stimulus conditions of R's firing. *Opacity*, and its opposite *transparency*, in the metaphorical sense at issue in Shea's argument, bear on the degree to which explanation promotes understanding in intentional agents, in the scientific community, in students reading textbooks, and so on. Transparency in an explanation is certainly a theoretical virtue, but it is a pragmatic virtue characterizable only in intentional terms.

Moreover, looking at Shea's and Neander's accounts side by side makes clear that this pragmatic virtue is evaluable only relative to a choice of explanatory target. Shea assumes that the explanation of how the system was stabilized by evolution carries the day. His preferred content—*nutritious flying object*—makes for a more transparent explanation of *that* target. Neander's account privileges a different explanatory project—providing a causal/mechanical explanation of the anuran visual mechanism, which is the primary concern of neuroethologists. A content attribution that captures the stimulus conditions of the mechanism makes for a more transparent explanation of *that* target, and so, as neuroscientists often do, they talk about the structures representing their stimulus conditions—that is, *moving wormlike stimulus (at a particular location)*.

Neander seems to acknowledge the relativity of content attribution to pragmatic interests when she says:

> The sign-stimulus for . . . [anuran] prey capture is a certain configuration of visible features [moving, wormlike]. To understand the relevant capacity is to understand how the normal [anuran] can recognize this configuration of visible features. The right content ascription in this case, *for these explanatory purposes*, concerns this configuration of visible features. (2017, 123, emphasis added)

To summarize: Both Shea and Neander appeal implicitly to broadly pragmatic considerations, but to considerations relevant to the different explanatory goals of different scientific communities. The different content attributions favored by these communities are motivated by their different explanatory concerns. So, despite appearances, Shea's and Neander's accounts are not in competition. Each isolates what may turn out to be components of a comprehensive naturalistic account of the anuran mind, but not one that will naturalize content in the advertised sense, inasmuch as the content attributed has a pragmatic rationale.

One might object that there is little evidence of such pragmatic choices in either the scholarly papers that scientists write for their peers or in textbooks and more popular works written for wider consumption. If content attribution is indeed pragmatically motivated, surely there should be explicit evidence of theorists' pragmatic choices. In fact, though, we shouldn't expect there to be explicit evidence of pragmatically motivated choices. When scientists write for each other, their pragmatic choices will normally not be made at the level of individual thinking. The explanatory interests that structure their inquiry and motivate such choices function as shared background, often not explicitly articulated by theorists. And this is similar for work targeted at a wider audience, where one would hardly expect to see the "sausage-making" involved in theory construction and development on display for the uninitiated. In both cases, the pragmatic choices underlying the assignment of cognitive contents will tend to be *invisible*. Especially when the goal of theorizing is to explain how a process works, it can look as if a tracking relation—the structure tracking its stimulus conditions—determines unique contents, naturalistically, all by itself. What the process was selected for, or even the function it fulfills in the larger economy of the organism, are not these theorists' concern, so alternative contents are not even in the running.

It might be pointed out that scientists make pragmatic choices in developing their theories all the time, not just in cognitive neuroscience.[14] They make choices that maximize the pragmatic virtues of intelligibility and theoretical simplicity. This is exactly what we should expect. So why, one might wonder, are pragmatically motivated choices in the *attribution of content* illegitimate? Of course, there are pragmatic choices all over science, but this fact should provide no comfort to a proponent of the naturalistic semantics project. *That* project explicitly calls for a non-semantic and non-intentional basis for content, and so prohibits appeal to the interests and intentions of agents in the determination of content. Pragmatic choices depend on such an appeal, if only implicitly, so they cannot play a role in a naturalistic account of content, as proponents of naturalized semantics have conceived of the project. However, as we shall see, the fact that pragmatic considerations are an ineliminable aspect of content attribution allows content to serve various important functions in cognitive scientific theories. It is a feature not a bug.

1.3.3 Content as Explanatory Gloss

Some of the structures posited in computational theories are assigned content in what we can think of as an *explanatory gloss* that accompanies the theory. Explanatory glosses are common in science; they serve various heuristic and pedagogical purposes. Niels Bohr glossed the atom as a planetary system. Bohr's planetary model was embraced by physicists when first introduced because it was successful in modeling the then salient behavior of atoms in terms of the quotidian and familiar. Electrons reside in different, discrete stationary states, which in the model are called *orbits*. The planets are held in orbit by gravitational force, electrons by the electrostatic attraction between electrons and protons. Moreover, it is a truism that the postulates of a theory can be understood more readily when embedded in a familiar model than when given a purely mathematical exposition. Maxwell's efforts to represent Faraday's lines of force—to *gloss* them, as it were—by the flow of liquid through tubes was an attempt to make intelligible a purely formal exposition of unfamiliar phenomena (in this case, electromagnetic phenomena) by appeal to systems governed by the laws of fluid mechanics, which had the status of familiar principles. Typically, the gloss—in our examples, Bohr's planetary model and Maxwell's appeal to liquid flow—is not considered part of the theory itself. Rather, it is a device that serves various heuristic and pedagogical purposes.

What I will call an *intentional* gloss—a particular type of explanatory gloss—construes the internal structures posited in the computational theory as *representing* external objects and properties. So, for example, a particular structure posited in a computational theory of vision might be characterized as representing *edges* in a visual scene. This representational content, and so the pragmatic elements that help determine the content, in particular, the collective goals and intentions of theorists, are not in what I will call the computational *theory proper*, but are isolated—or "quarantined," to use Mark Sprevak's (2013) useful expression—in the intentional gloss. Distinguishing between the computational theory proper and an intentional gloss that accompanies it preserves the naturalistic prospects of the computational theory itself. I will have more to say about naturalism in chapter 2.

Let me spell out the distinction between the theory proper and the intentional gloss more precisely. The *theory proper* of a computational account of a cognitive capacity comprises a specification of:

1) the mathematical function(s) computed by the mechanism. For example, the visual system computes smoothing functions to eliminate noise, the motor control system may compute vector subtraction in enabling an organism to reach for an object in view, and so on. (Recall that the specification of the function determines mathematical content. The inputs and outputs of the mechanism represent the arguments and values of the function.);
2) the specific algorithms involved in the computation of the function(s);
3) the structures that the algorithm maintains;
4) the computational processes defined over these structures, and ultimately how these structures and processes are realized in neural matter (the latter specified by what I am calling the *realization function* (f_R) as described earlier in the chapter); and finally

5) the *ecological component* of the theory—general facts about the organism's normal environment that explain why computing the mathematical function(s) specified by (1) suffices, in that environment, for the successful exercise of the cognitive capacity to be explained.

Examples of the sort of facts cited in ecological component (5) include that *light comes from above* and that *objects are rigid in translation.* Recall that Ullman (1979) hypothesizes that the visual system recovers the 3D structure of moving objects by computing a function from three distinct views of four non-coplanar points on the object to the unique rigid configuration consistent with the points. Ullman's structure-from-motion mechanism succeeds in recovering 3D structure from this minimal data only because, in our world, most objects are rigid in translation. A visual mechanism adapted to a world where objects are typically squishy or jelly-like would need much more information to recover their structure, and so a computational theory that characterizes such a mechanism would specify a different function, one much more complicated than the one proposed by Ullman, and, correspondingly, would attribute different mathematical contents to the inputs and outputs of the mechanism. In general, the ecological component of the theory constrains and justifies the choice of function, and so the attribution of mathematical contents, specified in component (1). The ecological component helps explain why computing *that function* in the organism's normal environment is sufficient for the organism's manifest ecological success.

Ecological component (5) also specifies relevant correlations between distal property tokenings and tokenings of the internal structures posited in component (3) of the theory. For example, the tokening of an EDGE structure[15]—a structure characterized in non-semantic terms—may be correlated with (and caused by) a change in depth or surface orientation in the environment. The fact that the mechanism is thus responsive to relevant changes in the environment explains in part how the mechanism enables the organism to recover the 3D structure of the scene. Such correlations provide some objective support for the attribution of a distal content (*edge*) to the structure in an intentional gloss that accompanies the theory, but the structure is not characterized in the theory proper as *representing* anything. The content attribution typically requires additional pragmatic motivation, as I have been arguing.

Components (1) and (5) of the theory proper constitute David Marr's *computational level* of theory, specifying *what* the system does and *why* it is useful, respectively. Components (2)–(4) reside at Marr's *algorithmic* level, specifying *how* the function(s) characterized by component (1) is computed, though the realization function (f_R), which for convenience I have included in component (4), is part of Marr's *implementation* level.

Components (1) through (5)—which make up what I am calling the *theory proper*—constitute the commitments of a fully articulated computational cognitive theory. They fully explain the cognitive capacity that is the explanatory target of the theory. A realist attitude toward these elements is appropriate. The rest is intentional "gloss." Some of the structures specified by (3) may be glossed as *representations*—they may be assigned contents in the intentional gloss that accompanies the theory, and so serve as representational *vehicles*. Accordingly, at least some of the causal processes specified by (4) are glossed as *representational processes*.[16]

1.4 Functions of an Intentional Gloss

An intentional gloss, which attributes determinate cognitive contents to some of the structures posited by the theory proper, serves at least *three* important explanatory functions:

1. *The intentional gloss connects the theory proper with its intentionally characterized explanatory target.*

The primary function of an intentional gloss is to illustrate, in a perspicuous and concise way, how the computational theory addresses the intentionally characterized phenomena with which inquiry began and that it is the job of the theory to explain. The explanatory target is typically characterized in such terms as, for example, the organism's *knowing* the 3D structure of the scene, or (in the case of the desert ant) *knowing* its way back to the nest. Cognitive content is the "connective tissue" linking the sub-personal, causal processes posited in the theory and the manifest, personal-level, intentionally characterized capacity that is the theory's explanatory target. Accordingly, vision theorists assign *visual content*; they look to properties that can structure the light in appropriate ways, and so they characterize the structures they posit as representing light intensity values, changes in light intensity, and further downstream, changes in depth and surface orientation. Theorists of motor control characterize the structures they posit as representing positions of objects in nearby space and changes in body joint angles.

The assignment of such task-specific distal content is justified only if the theorist can explain how the posited structures are used by the system in ways that subserve the cognitive capacity in question. As I noted earlier, the computational theory proper can fully explain the interaction between the organism and the environment, and hence the organism's success, without adverting to cognitive content, but appeal to cognitive content can make the explanation salient, given the theorist's explanatory goals. The intentional gloss characterizes the interaction between the organism and its environment in terms of the former *representing*, and occasionally *misrepresenting*,[17] elements of the latter, but the claim that structures posited in the theory represent these distal elements is not what explains the organism's success. Rather, the representational construal is a *gloss* on the mathematical and causal processes characterized by the theory proper that are doing the real explanatory work.

2. *The intentional gloss selects what is important or salient in a complex causal process, given the theorist's explanatory goals.*

Consider once again the frog/fly example. The causal process that begins with a fly entering the frog's visual field and eventuates in the frog's tongue catching the fly will include particular proximal stimulations on the frog's retina, but the retinal pattern will differ from case to case, with no particular pattern relevant to the explanation of the general capacity. Recall that excluding particular retinal patterns from an internal structure's extension is a problem for information-theoretic semantic accounts, which ground the representation relation in causal relations. The problem requires a pragmatic solution: content attribution involves a pragmatic-based *selection* over the entire causal process. The signal from distal triggering event to the tokening of the internal structure contains information about bugs, frog food, and small, dark moving things, as well as proximal retinal patterns, but not all of this information is relevant to explaining how the mechanism underlying the frog's prey-recognition capacity works, which is the primary explanatory target of computational neuroscience. *That* explanatory task requires specifying the response profile of the relevant neural structures and the role that these structures play in subsequent processing. All of this is provided by what I have called the *theory proper*. As Neander's (2017) reconstruction of the neuroscience of the anuran toad's visual system makes clear, a description something like "moving wormlike stimulus" best captures the response profile of the cells whose firing plays the appropriate role in the prey-recognition process (that is, T5-2 cells in the optic tectum), and so theorists often describe these structures as *representing* a moving wormlike stimulus. In doing so, they select from the complex causal process what is most important given their explanatory goals and gloss the causal process in representational terms. If the theorist had had Shea's explanatory interest—that is, explaining how the system was stabilized by evolution—then a different content attribution, privileging a different aspect of the causal process, namely that it is typically initiated by the presence of anuran food, would have been appropriate.

A second example will underscore the point. The discovery in the 1970s of *place cells* in the rat hippocampus was crucial to characterizing the mechanism underlying spatial navigation in rats.[18] Place cells are a parade case for robust representational realism, but such a reading overinterprets theorists' casual talk. Place cells respond to particular regions of the local environment. Neuroscientists gloss these causal processes using representation talk. The cells respond to particular locations; neuroscientists gloss their firing as *representing* those locations. "Head-direction" cells respond to head direction and are said to *represent* head direction. As before, the theorist's job is to specify the conditions to which the cell's firing is responsive and determine its role in controlling subsequent behavior. Talk of the cell's firing *representing* its distal stimulus conditions is a convenience that adds nothing of theoretical significance. The primary function of content attribution here is characterizing the internal structures in a way that highlights their causal role in the processes underlying the rat's navigational capacity. The representational talk is a *gloss* on these causal processes. Once cell firings are said to represent their distal stimulus conditions—here, locations in the nearby space—and assuming that place cells compose in an appropriate way, it is but a small step to construe the array of firings in the hippocampus as a *map* of the local environment that the rat uses in navigating, and so theorists sometimes gloss the array in these terms. We are familiar with the notion of a map from our ordinary lives; we use maps all the time to get around. But the idea of a subject using a map to navigate is just a helpful analogy; it functions in much the same way as Bohr's planetary model helped physicists understand atomic processes by glossing the discrete states inhabited by electrons as *orbits*. Elements in the neural array straightforwardly cause neural mechanisms in the rat to respond in appropriate ways, given the rat's location and its immediate goals. We can describe the process as "using a map" if we like—and to a point it is helpful to do so—but in any case, the user of "the map" is not the rat; it is a mechanism in the rat's brain.[19,20]

3. *Cognitive contents can serve as a placeholder for yet-to-be-specified computational structures and processes.*

Cognitive content can play an important role in the early stages of theory development. It often serves as a *placeholder* in an incompletely developed computational theory, prior to the specification of the structures and processes that make up what I am calling the "theory proper," guiding the search for the mechanisms underlying the cognitive capacity. For example, a visual theorist may hypothesize that cells in primary visual cortex respond to, and so *represent*, edges, and then go on to investigate which structures are involved and attempt to characterize precisely their distal stimulus conditions. Once the theoretical heavy lifting is done, the theorist may revise the initial content attribution—the representational characterization—to more accurately *gloss* the causal process characterized by the theory proper, as described in (2) above.

For a central example of representational content serving as a placeholder for a yet-to-be-developed scientific theory, consider our commonsense practice of predicting and explaining behavior by appeal to subjects' propositional attitudes (beliefs, desires, intentions, and so on). At some point, cognitive science may be in a position to specify the underlying neural realizations of various propositional attitudes and characterize the causal processes that eventuate in behavior. In the meantime, we can think and talk about the attitudes and describe their interactions only by reference to their contents. That said, there is no reason to assume that content is only a temporary expedient, that the neural/scientific story will replace commonsense psychology in our everyday use. Commonsense psychological explanation is the topic of chapter 3. I will argue there that the popular view that commonsense psychology is "proto-cognitive science" is mistaken. A computational theory that specified the neural structures and processes that realize the attitudes would be unlikely to subserve commonsense psychology's various functions particularly well.

1.5 Summary, and an Objection

Let me summarize the key components of the account of representation in computational cognitive science defended here:

1. A fully articulated computational theory specifies precisely the function computed by a cognitive mechanism, and in so doing characterizes the inputs and outputs of the mechanism as representing the arguments and values of the function. I call these *mathematical contents.*
2. It posits structures that play specified causal roles in cognitive processes. A realist attitude toward these structures is appropriate.
3. Some of these structures may be assigned cognitive content in an intentional gloss, where they function as representational vehicles. But the structures serving as vehicles in the gloss are not *fictions*; the attributed content is a gloss on real, causally efficacious structures. Place cells, for example, are typically glossed as representing locations in the immediate space, but they are not fictions.
4. The intentional gloss serves the various heuristic purposes I have described.

Points (1)–(4) constitute what I am calling a *deflationary view* of representation. Most importantly, no substantive representation relation holds between the structures playing the role of vehicles and the objects and properties they are taken in the gloss to represent. Moreover, the deflationary view emphasizes the role of pragmatic considerations in the attribution of representational content.

It has been suggested that the deflationary view is *unstable,* at risk of collapsing into either robust realism about representation on the one hand or eliminativism or some other form of antirealism on the other.[21] The worry about such a collapse is unfounded; it is based, I believe, on a failure to distinguish sufficiently between the representational vehicle and the content that the vehicle carries, a widely accepted distinction that is central to understanding my deflationary view. The structures that are characterized by what I am calling the *realization function* (f_R)—the bottom mapping in figure 1.2—and that when assigned contents function as representational vehicles, are *concreta*, as real as atoms, genes, and tables and chairs. Contents, on the other hand, are *abstracta*. This should be uncontroversial, so when robust realists claim that contents are as much a part of reality as other posits made by scientific theories, they can't mean quite what they say. My deflationary view offers a plausible way of salvaging much of what they are saying.[22]

It is, then, too simplistic to frame the issue, as Ramsey (2020) and others do, as an issue about whether one should be realist about the posits of our best scientific theories. In general, I am sympathetic to scientific realism, but, as I said, given a specification of an empirical theory, there is always a question about which aspects of the specification one should be realist about. Figuring that out requires not only close examination of the theory but also, often, stepping back to consider the norms and conventions that structure inquiry in the theoretical community more generally. Theories seldom wear their ontological commitments on their sleeves. Theorists themselves can disagree, never mind philosophers with different general metaphysical proclivities.

What is distinctive about the deflationary view of representation is that what I am calling *cognitive content*—ordinary distal content—is determined pragmatically by the theorist's explanatory goals and interests. Assessing the representational commitments of a particular computational model is straightforward on the account I have defended here. Assuming that the goal is to explain a cognitive capacity, then if the model posits structures that are assigned contents, then it is representationalist; if it does not, then it is not. One need not take a stand on the status of cognitive content to assess whether a particular computational cognitive model is representationalist or not.

Returning to the instability charge, Buchanan and Grzynkowski (2022) suggest that my brand of deflationism may be seen as the first step in an argument leading to eliminativism, in particular, to a wholesale rejection of any commitment to representation in cognitive science. That may be, but that is not *my* agenda. I have sketched a viable alternative to eliminativism, clearly indicating where one should get off the bus. Anyone skeptical of the robustly realist assumption that there is a substantive, naturalistically privileged representation relation but disinclined to eliminativism will find themselves in hospitable territory. I will have more to say about the contrast with eliminativism in chapter 2.

2 Reconstructing Intentionality

Brian Cantwell Smith (2002, 24) has suggested that to serve as a foundation for theorizing about the mind, theories of cognition must discharge their intellectual debts; in particular, they must discharge any commitment to *meaning* and *intentionality*. Robert Cummins (1989) makes a similar point when, referencing Dennett (1978), he says:

> To a large extent empirical theories of cognition can and do take the notion of mental content as an explanatory primitive. But this is a kind of explanatory loan . . . if . . . no account of the nature of the (mental) representation relation can be given that is consistent with the empirical theory that assumes it, then . . . that empirical theory must be regarded as ill-founded, and hence as a less than adequate response to the drive for the kind of thorough intellectual understanding that motivates scientific theory in the first place. (1989, 1–2)

At the same time, a plausible adequacy condition on a theory that claims to explain human cognition is that it maintain contact with the way that we see ourselves. There is a *prima facie* tension between these two goals: if cognition must be explained without appeal to meaning and intentionality, then what remains of the idea, central to our commonsense conception, that we *represent* the world? This chapter aims to resolve that tension. I argue that in offering an explanatory account of the mechanisms responsible for our cognitive capacities, computational cognitive science promises to satisfy a feasible agenda for naturalizing intentionality, though one quite different from the so-called naturalization project typically endorsed by philosophers of mind, particularly those of a robust representational realist bent.[1] The naturalization of intentionality so reconceived—what I call *reconstructing intentionality*[2]—is something that computational cognitive science has a good chance of actually delivering.

Here is the plan for the chapter. I begin by considering two proposals for discharging apparent commitment to meaning and intentionality—the reductive program advocated by proponents of the naturalization project and the deflationary program I develop in chapter 1. I argue that the deflationary program avoids the serious problems that beset the naturalization project and better characterizes actual practice in computational cognitive science. In section 2, I defend the deflationary program against two prominent objections to computationalism. Section 3 sketches an alternative agenda for naturalizing intentionality, one that I argue computational cognitive science, properly construed, is well placed to satisfy. Finally, in section 4, I motivate the idea that an explanatory theory of cognition should maintain contact with our commonsense conception of ourselves as representing the world, and I explain how computational cognitive science, by making pragmatic use of an appropriate intentional gloss, achieves this goal. I argue that an explanatory gloss accompanying a theory and serving various expository and heuristic functions is a general feature of scientific theorizing and not an ad hoc feature of the deflationary account of computational cognitive science defended here.

2.1 Discharging Debts to *Meaning* and *Intentionality*—a Well-Worn Proposal and a More Modest Alternative

Computational cognitive psychology and computational neuroscience are the branches of cognitive science that aim to provide a foundation for the scientific study of the mind. Theorists working within the computational program aim to *fully* explain the cognitive capacities that are the targets of their theorizing. The requirement that a foundational theory of mind discharge any commitment to meaning and intentionality implies that *representation* cannot serve as an explanatory primitive in such a theory.[3]

I will use "discharge" as an umbrella term that covers a range of possibilities for repaying the apparent debt that a theory incurs in virtue of appealing to such notions as *representation* and *content*. One option is simply to deny that computational theories make a genuine appeal to representation, holding that the commitment is merely apparent. *Fictionalism* and *eliminativism* are two versions of this strategy. Fictionalists claim that talk of neural representation does not denote actual structures in the brain but is simply a useful fiction serving various purposes.[4] I argued in chapter 1 that theories that purport to explain cognitive capacities often posit structures that, when interpreted, count as representational vehicles. These structures play essential roles in causal explanations of cognitive processes. Place cells are an example. Fictionalism is therefore not supported by actual practice. Eliminativists, while granting that cognitive theorists often use representational talk, argue that appeal to representation serves no useful purpose and so such talk should be eliminated. I will set eliminativism aside for now, returning to it at the end of the chapter.[5]

That leaves at least two ways that a theory might discharge a commitment to meaning and intentionality:[6]

1. A *naturalistic reduction* of meaning/intentionality—the specification of non-semantic and non-intentional sufficient conditions for states of mind to represent aspects of the world.
2. A *deflationary construal* of representational talk, according to which appeal to meaning or content plays only a heuristic role in a computational theory, rather than an individuative or otherwise substantive role.

I will discuss each proposal in turn.

2.1.1 Discharging Debts to Meaning: A Naturalistic Reduction of Intentionality

The so-called *naturalistic semantics project* (or simply *the naturalization project*) is motivated in part by the conviction that intentionality, while a real property of mental states, is not fundamental. Fodor, in a well-known passage, puts the point this way:

> I suppose that sooner or later the physicists will complete the catalogue they've been compiling of the ultimate and irreducible properties of things. When they do, the likes of *spin, charm,* and *charge* will perhaps appear upon their list. But *aboutness* surely won't; intentionality simply doesn't go that deep. It's hard to see, in face of this consideration, how one can be a realist about intentionality without also being, to some extent or other, a reductionist. If the semantic and the intentional are real properties of things, it must be in virtue of their identity with (or maybe supervenience on) properties that are themselves *neither* intentional *nor* semantic. If aboutness is real, it must be something else. (1987, 97)

The naturalization project is the attempt to specify what that "something else" might be. The specification of a semantics for mental states that satisfies the naturalistic constraint would demonstrate the dependence of intentional mental phenomena on phenomena that already yield to empirical investigation. This project has been pursued vigorously by philosophers for the last forty years,[7] and there is something of a consensus that no one has yet specified a genuinely non-intentional and non-semantic basis for intentionality, though work on the project continues apace.[8] There are a number of reasons why the traditional naturalization project might fail:

1. Fodor's claim notwithstanding, intentionality may be a primitive property of minds after all, as proponents of the *phenomenal intentionality research program* (PIRP) presume.[9] The idea here is that determinate content is grounded in a subject's phenomenal experience. Or it may be genuinely emergent out of physical and biological processes characterized by the sciences, and so beyond the scope of explanation grounded in these sciences. Either way, the hopes of computational theorists to fully explain human cognition would be dashed.
2. Even if there *are* non-semantic and non-intentional sufficient conditions for intentionality, the naturalistic reduction base may be too complex to be tractably specifiable, with the result that the naturalistic project would fail to yield an *explanation* of intentionality that would advance our understanding of the phenomenon.

Shea (2013) notes: "A naturalistic account of mental content must provide *illuminating* explanatory connections between representational content and properties that are non-semantic, non-mental and non-normative" (497, emphasis added). The naturalistic account should not leave us scratching our heads, wondering of the proposed reduction base what *that* has to do with content.

Fodor's (1990) *asymmetric dependency theory*, while not, of course, itself intractable, nonetheless illustrates the worry. The final version of the theory has four somewhat nonintuitive conditions that block some counterexamples but leave us wondering why satisfying those particular conditions should yield meaning. Recall from the discussion in chapter 1 the trouble that information-theoretic accounts have in excluding particular retinal patterns from the extension of an internal structure's tokening. In light of this "causal chains" problem, Fodor revises his notion of robustness: "The dependence of Cs on Bs is robust only if there are non-B-caused Cs" (118). So the account of robustness requires that there *actually be* heterogenous causes of the internal token, not simply that there could be. This adds an additional condition to the three conditions cited in Fodor's earlier (1987) account: some "X"s are caused by non-Xs (1990, 121). One might wonder, though, why meaning (and robustness) requires *actual* heterogeneous causes, rather than just the possibility of them. The possibility of heterogeneous causes (non-B caused Cs) is sufficient to allow for the possibility of misrepresentation, a requirement for genuine representation, but Fodor's additional requirement, while blocking some counterexamples to the original analysis, seems ad hoc, adding nothing illuminating to the account.

3. The phenomenon in question, viz., representation, may in fact be pragmatically determined, depending in part on the collective interests, goals, and intentions of communities of attributors, as I argued in chapter 1. If this is the case, then a full naturalistic reduction of intentionality would require the specification of non-semantic and non-intentional sufficient conditions for the determinate interests, goals, and intentions of attributors. Such a specification would be an

even more ambitious project than proponents of the naturalization project have envisioned. Even if there are such sufficient conditions for theorists' goals and intentions, they may not be tractably specifiable, and so, again, the reduction would fail to yield an explanation that would advance our understanding of intentionality.

One source of skepticism about the naturalization project is the worry that reference or aboutness is an entirely different kind of relation than the relations that non-basic, mid-level science typically traffics in—that is, *causal* relations. Reference, after all, is extremely precise. As Quine (1960) pointed out, it cuts between rabbits and temporal stages of rabbits; causation does not. Moreover, we can think and talk about things that are very far away in space and time, for example, GN-z11, a galaxy 13.4 billion light-years from Earth, the Big Bang, and abstract objects such as the square root of 2. It would be nice to have an account of how this is possible—perhaps somehow by construction out of spatially and temporally more local non-semantic and non-intentional processes, as proponents of the naturalization project must presume—but the scientific study of the mind cannot wait for such an account, and computational neuroscientists and psychologists engaged in explaining our cognitive capacities presume nothing of the sort.

Nicholas Shea, a proponent of the naturalization project, has suggested that cognitive science look to philosophy for the hoped-for reduction. He maintains that cognitive science takes semantic properties for granted, "offer[ing] no settled view about what makes it the case that the representations relied on have the contents they do. The content question has been largely left to philosophy" (Shea 2013, 499). If Shea is right, then it would seem to follow that cognitive science cannot provide a *foundation* for theorizing about cognition. At best, it can supplement work done by philosophers. I suspect, though, that cognitive scientists would be quite surprised to hear that their theorizing requires supplementation by philosophers. Moreover, as I argued in chapter 1, the latest philosophical accounts of content, Neander (2017) and Shea (2018), appeal to pragmatic elements and hence fail to secure a naturalistic basis for the attribution of content. Nonetheless, there is a sense in which Shea *is* right. On the account of content attribution sketched in chapter 1, there is a sense in which computational theorists do take content for granted. Rather than reducing content to non-semantic or non-intentional properties or appealing to some naturalistic representation relation to ground their determinate content attributions, they *use* unreduced, pragmatically motivated content attributions to explicate (gloss) their theories and to serve the various heuristic functions I described in the previous chapter. In doing so, they help themselves to the ontology implicit in public language—that is, to *moving wormlike things* and to *edges*, rather than to temporal stages or undetached parts of such things.[10] But, *pace* Shea's suggestion, the naturalistic bona fides of computational cognitive science does not depend on philosophers finding the holy grail of a naturalistic content-determining relation.

Let me turn then to (2), a more modest proposal for discharging apparent commitment to meaning and intentionality.

2.1.2 Discharging Debts to Meaning: A Deflationary Alternative

Since there is little evidence that computational theorists themselves are interested in the traditional naturalization project of philosophers, I suggest that a deflationary construal of representational talk offers a more promising way of discharging the apparent commitment of computational theories to meaning and intentionality. On the deflationary account of representation defended in chapter 1, the determinacy of such distal content as *edge* or *moving wormlike thing*—what I call *cognitive content*—is secured only by implicit appeal to theorists' explanatory goals and intentions, but appeal to these pragmatic elements does not undermine the naturalistic credentials of a theory. Cognitive content, recall, is not in what I call the *theory proper* but rather is "quarantined" in a pragmatically motivated intentional gloss. The deflationary strategy explains why the problematic phenomena—meaning and intentionality—might appear to be on an ontological par with the structures and processes posited by the theory proper. They are not, but that is not to say that the gloss doesn't play an important role in making clear how the theory proper explains what it is intended to explain, and in a way that marries that explanation with our commonsense conception of ourselves and our cognitive capacities. The basic point here, which cannot be overemphasized, is: a computational theory's commitment to meaning and intentionality is *merely* apparent. In making pragmatic use of cognitive content, a computational theory incurs no explanatory debt that needs to be discharged and so no debt that it is the job of philosophers to discharge.

Computational theories, properly understood, do not traffic in cognitive content, such content being confined to the gloss. Since it is not in the theory proper, cognitive content is not in need of naturalizing. But computational theories do, I have argued, traffic in another sort of content, what I have called "mathematical content," which, it will be recalled from chapter 1, is determined by the specification of the function computed by a computational mechanism. The inputs to and outputs of a computational process represent the *arguments* and the *values*, respectively, of the function computed. This mathematical content *is* essential to the computational characterization of the mechanism: it specifies what the mechanism does. This sort of content, it might be thought, does stand in need of naturalization. As Facchin (2022) puts it, "The deflationary account accepts that a satisfactory account of representations must deliver naturalized contents. Yet, mathematical contents have not been naturalized, and I claim that it is very doubtful that they ever will" (2022, online preprint, 1).

In fact, mathematical content does not stand in need of naturalization; it too can be discharged.

The apparent debt of a computational theory to mathematical content is discharged by the theory, and we can see how by focusing again on our addition example (figure 2.1).

Recall that a physical system has the capacity to compute the addition function just in case there exists a mapping from physical states of the system to numbers, such that physical states related causally (p1, p2 → p3) are mapped to numbers <u>n</u>, <u>m</u>, and <u>n+m</u>, related as addends and sums. The mapping specifies the causal organization of the system underlying its capacity to produce behavior consistently interpretable as adding. The mapping itself discharges the debt to mathematical content. To say, for example, that the system adds 53 and 47 to get 100 is to say that when it is in the physical state interpreted under the mapping as 53 and in the physical state interpreted under the mapping as 47, it then goes into the physical state interpreted under the mapping as 100. This causal organization, assuming it holds across a sufficient range of the function, allows us to say that the system

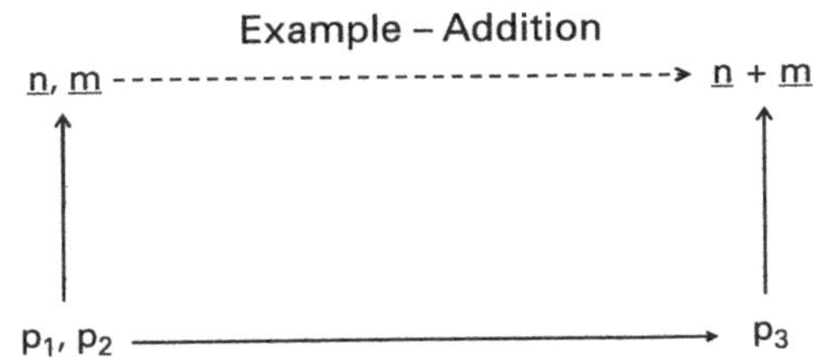

Figure 2.1
Example—an adder.

computes addition.[11] The mapping to the addition function provides a compact characterization of the causal structure responsible for the system's capacity to add. It predicts how the system will behave in a wide range of circumstances that go well beyond the observed behavior of the device. Importantly, there is no unreduced representation relation or content here.

In effect, the mapping of physical states to the arguments and values of the computed function in the computational explanation of a cognitive capacity *naturalizes* mathematical content, specifying the "relation" that supports the particular assignment of content to the system in non-semantic and non-intentional terms. The relation, of course, is simply given by the mapping. The mapping does not pick out a substantive relation that states of the device bear to mathematical objects, something that one might be tempted to call *the representational relation.* Rather, as I've said, the mapping specifies the causal organization underlying the target cognitive capacity. To have the causal structure specified by the mapping—the relational structure that holds among physical states p_i—is all there is for a physical system to have the capacity to compute the specified function. The capacity may or may not be exercised, depending on whether the system receives as input states that under the mapping represent the arguments of the function. (See the next section for more discussion of this point.) Computing the function is all there is for states of the system to represent the arguments and values of the function.

Rather than construing the mapping as naturalizing mathematical content, we might say that mathematical content, as I characterize it, does not *need* naturalizing, any more than physical measurement, which involves a mapping between physical magnitudes (for example, mass, length, or temperature) and elements of a numerical scale, stands in need of naturalization. The abstract relational structure shared by the empirical domain and the numerical scale underlies and supports the mapping; there is no substantive relation between a measured state of the system, say, having a particular mass, and its numerical "representative," say, the number 7 (grams). It is similar for physical states p_i (or s_i in figure 1.2) and the mathematical objects to which they are mapped. Whether we construe the mapping as naturalizing mathematical content or not, the apparent commitment of a computational theory to mathematical content is discharged.

In summary, then, computational theories, properly construed according to the deflationary account, discharge their apparent commitments to meaning and intentionality and so are apt to serve as a naturalistic foundation for the study of cognition. Cognitive content is confined to a heuristic gloss and so does not play a substantive or essential role in the computational theory proper. The fact that its attribution depends on various pragmatic considerations, therefore, does not impugn the naturalistic credentials of the theory. Mathematical content is a consequence of the specification of the function computed by the system. The mapping that specifies this content picks out the relevant causal organization of the system without positing any substantive relation between states of the system and the mathematical objects used to specify that organization, and without any appeal to intentional or semantic notions. Mathematical content is already naturalistic.

In general, though, to discharge a (apparent) commitment to intentionality is not thereby to naturalize intentionality. Fictionalist and eliminativist proposals, for example, do not naturalize intentionality. I sketch below an agenda for naturalizing intentionality quite different from the reductive program offered by traditional naturalistic semantics, an agenda that computational cognitive science, construed along deflationist lines, has a good chance of satisfying. Before that, I address two prominent objections to the computationalist program itself, however it is construed. The deflationary account has the resources to definitively answer these objections.

2.2 Two Objections to Computationalism and the Deflationist Response

2.2.1 Mathematical Content Is Not Unique: The Problem of Unintended Interpretations

I have argued that mathematical content is determined by the specification of the function computed by a computational mechanism. A common objection holds that a physical system that computes a given function, say addition, can be interpreted as computing other functions as well. The causal transitions among physical states p_i (depicted in figure 2.1, or more perspicuously, abstractions over these states functioning as vehicles s_i in figure 1.2) can be mapped to other functions isomorphic to addition for at least some part of its range. For example, Facchin (2022) argues that, as I have characterized it, the adder can be interpreted as computing a function from the first nine US presidents to the set of presidents from John Adams, the second president, to Ulysses S. Grant, the eighteenth president. But then what do the physical states of the system *really* represent—numbers (related as addends and sums) or presidents? Inasmuch as mathematical content is determined by the function computed by the mechanism, Facchin concludes, it is radically indeterminate.

Before responding directly to the indeterminacy challenge, I would like to take a step back and remind the reader what computational theories of cognitive capacities of physical systems are supposed to do. Their job is to characterize the mechanism(s) responsible for the target capacity, in particular, to characterize the mechanism's causal organization. There may be multiple ways to do this (neuroanatomy and neurophysiology, for example, have their own methods) but computational theories do it by deploying a particular strategy—they construe the mechanism as a computer (or collection of computers). This strategy involves specifying the function(s) computed by the system, explaining how the specified function(s) is computed, and explaining how computing that function(s) enables the successful execution of the target capacity in the subject's normal environment. All of this is done by what I am calling the computational *theory proper*. It's a tall order; developing an adequate account is a considerable theoretical achievement.

Facchin is right that a physical system may be interpretable as computing any number of other functions. But there is no reason to think that an arbitrary function specification will support an explanation of the target capacity. Consider, for example, Facchin's function on presidents. How would computing that function explain a system's manifest capacity to add? Does it provide a perspicuous characterization of the system's activity that contributes to our understanding of the target capacity? An interpretation of the adder as a device that maps individual presidents into sets of presidents throws no light on the cognitive capacity that the theory purports to explain.

We can identify general constraints on the function specification—what in Egan (2017) I call the *function-theoretic characterization* of the mechanism.[12] The mathematical functions deployed in computational models are typically well understood independently of their use in such models. Laplacian of Gaussian filters, fast Fourier transforms, vector subtraction, and so on are standard items in the applied mathematician's toolbox. To apply one of these tools to a biological system—to subsume the system under the mathematical description—makes sense of what might otherwise be a heterogeneous collection of input–output pairs. ("I see what it's doing . . . it's an integrator!") Bizarre functions such as Facchin's function on presidents will not have this feature. Moreover, computational theorists typically have at their fingertips various algorithms for computing these independently well-understood functions, algorithms that serve as preliminary hypotheses for *how* the mechanism computes the function. Importantly, we have some idea how devices capable of computing such functions can do certain things, perform certain actions, and so on. The theorist will select the function that, given these constraints and evidence concerning both behavioral performance and neural circuitry, does the best job of characterizing the causal processes responsible for the capacity to be explained. We can call this the *canonical specification of the function computed.* The inputs and outputs of the mechanism are thereby interpreted as representing the arguments and values, respectively, of this function. We can say that this is what the system *really* represents if we like, but the attribution doesn't carry the metaphysical weight that robust representational realists (see chapter 1) would assume. As I have stressed, it doesn't depend on any substantive relation between states of the system and the mathematical objects that constitute the arguments and values of the function.

In criticizing my deflationary account Facchin goes on to say, "A suitably robust account of physical computation should enable us to say that any physical system S computes few (ideally one) function F" (2022, 9). But I am not offering a metaphysical account of physical computation; rather, my goal is to explain the role that appeal to computation plays in actual neuroscientific practice. Facchin fails to appreciate the role that mathematical content plays in the deflationary account, which, as I have stressed, is to characterize the causal organization underlying the system's exercise of the cognitive capacity to be explained. In effect, computational theories *model* the relevant causal processes as mathematical processes.

When the role of mathematical content in the deflationary account is properly understood, then Facchin's objection can be seen not so much as an objection to my notion of mathematical content, but rather as an objection to the very idea that a computational characterization of a physical system—a characterization of the system as computing a specified function (or set of functions)—can have genuine explanatory purchase. But what else could it be to construe a physical system—natural or artifactual—as a *computer*? The argument from the existence of unintended interpretations is a well-worn objection to computationalism, but it is ineffectual. A given computational characterization is subject to criticism if it fails to predict the system's behavior, or behavioral or other evidence suggests that it fails to characterize the causal organization responsible for the target capacity, or it fails to satisfy reasonable pragmatic constraints (the function is too hard to understand, it is not tractably specifiable, and so on), but its explanatory force is not undermined simply by pointing to the existence of unintended interpretations.

A qualification is in order at this point. What I am calling the "canonical specification" often gives only a rough approximation of the function computed. The initial characterization is likely to be selected from the computational theorist's toolbox, but these initial hypotheses will often be modified in light of new behavioral data, sometimes to the point where eventually the device is said to compute a function sufficiently different from the well-understood function with which the theorist began. The computational model of natural language processing developed by Marcus (1980) is a case in point: the proposed model is an augmentation of a standard LR(k) parser of the sort that one might encounter in a graduate-level course in parsing theory. The augmentations are dictated by observed features of human linguistic competence, and the resulting model differs in significant ways from the initial function-theoretic characterization. But this qualification reinforces the general point that it is the *causal relations* modeled by the mathematical functional relations that count and that the theorist is aiming to characterize; the function specification will be evaluated by how well it does that job.[13]

Before concluding the discussion of mathematical content, let me forestall a possible confusion. To say that the specification of the function computed, and the mathematical content so determined, characterize the causal organization underlying the system's cognitive capacity is *not* to say that content itself makes a causal difference to the behavior of the system. Content—mathematical or cognitive—*never* makes a causal difference to behavior. Rather, content is used by theorists to specify the system's efficacious properties, in particular, relevant aspects of its causal organization. That is the role of mathematical content in computational models.[14]

But one must be careful here. Acknowledging that mathematical content is not itself efficacious in my deflationary account, Hutto and Myin (2020) mistakenly conclude that computational explanation is non-causal: "The trouble with assuming that computational explanations are non-causal is that it becomes unclear what is gained by holding that cognitive systems actually make any contentful calculations at all" (96).

But of course, the *calculations* characterized by the mapping to mathematical content are physically realized processes that cause the system's behavior. Mathematical content, like all content, is itself causally inert, but the processes responsible for the cognitive capacities that it specifies are not at all inert. Consider once again figure 2.1. The bottom span of the diagram depicts causal transitions among physical states. The top span depicts a formal relation, mapping the arguments of the specified function into its values. As I have said, the causal relations underpinning cognition are modeled as formal or mathematical relations in computational theories, but they are no less causal for that.

2.2.2 Computational Characterization Is Trivial: The Problem of Unintended Realizations

The argument from unintended interpretations of computationally characterized systems is intended to show that computational "explanation" is trivial. Interpretations are cheap; how can they be genuinely explanatory? A distinct argument appeals to the existence of unintended *realizations* of computational characterizations to argue for the same conclusion. Hilary Putnam (1988) and John Searle (1991) famously argued that rocks, walls, indeed all physical systems turn out to be computers, because their physical states can be mapped to computational descriptions. If all there is to computing a function is the existence of a mapping from physical states of the system to the arguments and values of the function, then, they insist, computational "explanation" is utterly trivial; even the coffee cup in front of me computes addition in this sense.

My response here will be brief. I have no interest in attempting to identify additional constraints that together with the mapping condition specify sufficient conditions for a physical system to *really* compute a function.[15] As I have said, I am not offering a metaphysical account of physical computation. Rather, my goal is to motivate the idea that computational characterizations—in particular, the canonical specification of a function computed by the system—when appealed to in explanations of cognitive capacities, are not trivial but can be genuinely explanatory.

I have argued that the mapping to an independently well-understood mathematical function provides a compact specification of the causal organization responsible for a physical system's exercise of a cognitive capacity. So, for example, the mapping depicted in figure 2.1 specifies the causal organization of a system underlying its capacity to produce behavior consistently interpretable as adding. The requirement that the physical states specified at the lower level of the mapping be causally related will eliminate many spurious mappings. A complete explanation of the capacity would identify the physical states, interpreted as the arguments of the computed function, that cause the physical states that are interpreted as the values of the function. We can say that for a physical system to have the capacity to compute a given function, it must have a causal organization that can be mapped, in the direct way depicted in figure 2.1, to the function. In other words, causal state-transitions of the system mirror the formal dependency relations among the arguments and values of the function. The possibility of unintended realizations—say, a rock or a coffee cup happening to have the specified causal organization—has no implications for explanatory practice in cognitive science. In particular, it does not make computational explanation trivial. To see why, we need to focus on the *context* in which computational explanations are typically offered.

Theorists deploy computational descriptions to explain the manifest cognitive capacities of physical systems such as ourselves, capacities such as understanding and producing speech, recovering the three-dimensional layout of the scene, adding, and so on. Rocks, walls, and coffee cups have no manifest cognitive capacities. With the target capacity in mind, the theorist hypothesizes that the system computes a well-defined function and spells out how computing the value of the function would explain the system's observed success at the cognitive task. To cite just one example, described in chapter 1, according to David Marr's account of early vision, computing the Laplacian of a Gaussian of the retinal array produces a smoothed output that enables the detection of sharp discontinuities in intensity gradients across the retina, and hence the detection of significant boundaries (edges) in the scene. In other words, the computational description is justified by reference to the use to which the computation is put in the exercise of a manifest cognitive capacity. As I said earlier, the theorist is thereby committed to the system actually having the causal organization specified by the mapping. That is hardly trivial. It is not relevant to the explanation that there may be *other* physical systems with the same abstract causal organization, which, *if they were suitably hooked up to other internal mechanisms and situated in the right environment* (and these, it should be emphasized, are substantial requirements!) would enable the larger system to execute the cognitive task in question, say, detecting edges. Similarly, if a rock had a causal organization specified by the type of mapping to the addition function depicted in figure 2.1, and we could interface with it in such a way that we could put it into the relevant physical states and detect the relevant physical changes, as we can interact with a hand calculator, then we could use the rock as an adder. That would certainly be surprising, but this possibility does nothing to undermine the cogency of the computational explanation of the hand calculator. Physical systems like ourselves, who have lots of cognitive capacities, will satisfy many such mappings, each specifying the distinct causal organization underlying a specific cognitive capacity.

To summarize: The deflationary view has the resources to handle these two prominent objections to computationalism. In rejecting the idea that representational talk requires a substantive representation relation holding between structures posited in the theory and what they are construed as representing, and focusing rather on the uses to which the mapping that characterizes a computational system is put in the theory, the deflationary account secures the explanatory potential of computationalism. The possibility of unintended interpretations and unintended realizations is simply irrelevant to that potential.

The common idea underlying my response to the two triviality objections is that a computational theory is to be evaluated by how well it does its job of explaining a *manifest* cognitive capacity of a physical system. The possibility of unintended interpretations implies that there may be different ways of characterizing the relevant causal organization, and hence more than one way of doing that job. It is possible that two or more theories may satisfy the empirical and pragmatic constraints equally well, just as there may be two or more grammars that characterize a speaker/hearer's linguistic competence equally well, with nothing to choose between them. But such an embarrassment of riches should not be seen as undermining the explanatory bona fides of computational practice. As Chomsky has noted, it is hard enough for a theorist to come up with one adequate theory. The existence of unintended realizations has even less bearing on the explanatory potential of the computational program. The possibility of a rock's realizing the addition function does nothing to undermine a computational explanation of my hand calculator's capacity to add.

2.3 An Alternative Proposal for Naturalizing Intentionality

The deflationary view preserves the naturalistic prospects of the computational program by discharging any appeal to meaning and intentionality, but I have not said anything substantive about how, if at all, the phenomenon that we call "intentionality" might actually be naturalized. This, of course, is the aspiration of the traditional naturalistic project. I turn to this issue now. I recommend an alternative proposal that foundational theories of the mind are more likely to satisfy. I call the alternative *reconstructing intentionality*.

Naturalizing a phenomenon is a philosopher's notion; the notion plays no role in the sciences themselves. The so-called naturalistic semantics project has commandeered the idea of naturalizing intentionality in the wake of seminal work by Dretske (1981), Millikan (1984), Fodor (1987), and others, where naturalizing intentionality has come to be identified with the task of specifying non-intentional and non-semantic sufficient conditions for a mental state to have its particular content. Given the emerging consensus that the project has not succeeded, I urge that we not concede the label "naturalizing intentionality" to the naturalistic semantics project. It is time to consider other options.

I propose that naturalizing a phenomenon P should be understood simply as *naturalistically explaining* P. This core notion needs some unpacking. The philosophical literature distinguishes two kinds of naturalism: (1) *ontological* (or *metaphysical*) *naturalism*, which holds that there are no extra-physical (in some vague sense of "physical") objects, properties, and processes; and (2) *methodological naturalism*, which is not really a thesis at all but rather a commitment to apply empirical, scientific methods to the study of the phenomenon in question, in our case, mental phenomena.[16] However the constraints imposed by ontological naturalism are to be precisely understood, discharging any debts to meaning and intentionality, and construing mental processes as causal, mechanical processes, as computational cognitive science promises to do, will satisfy them. Methodological naturalism is also rather vague, but it requires, at a minimum, a commitment to objective methods of inquiry.[17] I have argued that the naturalistic semantics project is not consonant with actual practice in the cognitive sciences, importing constraints from elsewhere, notably from philosophy. The upshot is that the naturalistic semantics project, at least arguably, violates methodological naturalism.

So let us consider other options for how to understand "naturalizing intentionality." Andy Clark makes the following promising suggestion:

> To naturalize intentionality . . . "all" we need do is display the mechanisms by which . . . ongoing viability-preserving engagements [between organism and world] are enabled, and make intelligible that such mechanisms can deliver the rich and varied grip upon the world that we humans enjoy. (Clark 2015, 7)

Taking up Clark's suggestion,[18] I propose that we understand naturalizing intentionality as requiring an explanatory account of the mechanisms responsible for the processes that we pretheoretically refer to as *representational* processes, processes that secure our "rich and varied grip on the world." The proposal better describes actual practice in computational cognitive science than do philosophers' conceptions of the naturalistic semantics project, and so it satisfies methodological naturalism.

I argued in chapter 1 that providing an explanatory account of the mechanisms responsible for our cognitive capacities is what the computational theory proper aims to do. Let me review very briefly the main features of the account. The theory proper comprises a specification of (1) the mathematical function(s) computed by the mechanism, (2) the specific algorithms involved in the computation of the function(s), (3) the structures that the algorithm maintains, (4) the computational processes defined over these structures, and (5) the ecological component that specifies general facts about the organism's normal environment that explain why computing the mathematical function specified by element (1) of the theory suffices in that environment for the successful exercise of the cognitive capacity to be explained.

As I have explained in this chapter, the point of (1), the specification of the function computed, is to characterize the causal organization responsible for the cognitive capacity. Ecological component (5), then, explains why having *that* causal organization, in the subject's normal environment, is sufficient for the successful exercise of the capacity. Success is construed as success of the organism, given its goals, and is always evaluated relative to the organism's normal environment.

The theory proper provides an explanatory account of the mechanisms responsible for our cognitive capacities, including capacities that we pretheoretically describe as representational capacities. It does so by characterizing the posited structures and processes in non-semantic and non-intentional terms.[19] Some of the causal processes specified by element (4) of the theory may be characterized as representational processes in an explanatory gloss that accompanies the theory, but these should not be construed literally as instantiating *the representation relation*, where that entails that they have determinate accuracy or truth conditions. The theory will cite robust correlations between tokenings of internal structures specified by element (3) of the theory, for example, EDGE,[20] and distal states of affairs such as *object boundaries* in the scene, but the organism's "grip on the world"—what we can think of as the directedness of the subject toward relevant aspects of the world, in this case, its three-dimensional structure—is maintained by the causal relations underlying these correlations and explained in noncontentful terms. The specified internal structures are not characterized (in the theory) as *representing* edges in the world, or any other distal object or property. This directedness of the subject toward the world should *not* be construed as intentionality. The theory explains how the organism "gets a grip" on the 3D structure of the world, enabling its successful engagement with the world, but it doesn't deliver anything fine-grained enough to underwrite the assignment of determinate *edge* contents, as opposed to, say, *edge stage* or *undetached edge part*, or *fly*, as opposed to *frog food* or *small, dark moving thing*. Such determinate contents are motivated in part by pragmatic considerations and attributed only in the intentional gloss.

Similarly, a subject's successful reaching and pointing behavior, supported and explained by the sort of causal processes characterized by Shadmehr and Wise's (2005) computational account of motor control (described briefly in chapter 1) is directed at objects and locations in the subject's nearby space. The Tunisian desert ant's successful homing behavior, supported and explained by the causal processes characterized by Gallistel's (1990) computational account of the ant's navigational capacities, is directed at the nest and the intervening space. Importantly, *directedness*, as I understand it, is a relation between specified activities of the organism and aspects of the world. It is *not* a relation between internal states or structures posited in the theory proper and distal entities or property tokenings that these structures may be construed as representing in an appropriate intentional gloss.

One might object that since the notion of *directedness of the subject toward relevant aspects of the world* will not support determinate content and correctness conditions, the deflationary account falls short of naturalizing intentionality. If we concede the label "naturalizing intentionality" to proponents of the naturalistic semantics project, which requires sufficient non-intentional and non-semantic conditions for determinate content, then the objection is apt. But computational cognitive science, I have argued, does not aim to *reduce* intentionality. Rather, the theory proper, together with an intentional gloss that accompanies it, enables a *reconstruction* of intentionality—that is, of what we think of as full-blooded representation. Let me spell this out. I suggested that we understand naturalizing intentionality as requiring an explanatory account of the mechanisms responsible for the processes that we pretheoretically refer to as *representational* processes, processes that secure our "rich and varied grip on the world" (Clark 2015, 7). I have argued that this is precisely what computational psychology and computational neuroscience promise to provide. In doing so, these sciences characterize a sense in which the subject's activity is *directed at* relevant aspects of the world. Directedness is not an *intentional* relation; it doesn't yield determinate content and correctness conditions. However, when an account of the causal processes that support directedness, given by the computational theory proper, is supplemented by an appropriate intentional gloss, we do get determinate content and correctness conditions—that is, we get full-blooded representation. We get what I am calling a *reconstruction* of intentionality. If, as I have been at pains to argue, content attribution is (in part) pragmatically determined by theorists' explanatory goals and interests, then this is all we can reasonably expect by way of a naturalistic explanation of intentionality. Moreover, it is all we should want.

This reconstruction of intentionality is a crucial element in *maintaining contact with the way that we see ourselves*, the second adequacy condition I mentioned at the outset on a theory that purports to explain cognition. I turn to this issue now.

2.4 Maintaining Contact with the Way We See Ourselves

The idea of an explanatory gloss accompanying a theory and serving various expository and heuristic purposes is familiar in the philosophy of science. Stephan Hartmann (1999) appeals to a *story*, which he claims is "of utmost importance in almost all branches of actual scientific practice" (326).[21] He discusses in detail the case of Hadron physics, the branch of physics that investigates the structure and interactions of subnuclear particles such as protons, neutrons, and pions, but the idea, he emphasizes, is general:

> A story is a narrative told around the formalism of the model. It is neither a deductive consequence of the model nor of the underlying theory. It is, however, inspired by the underlying theory (if there is one) . . . . [T]he story fits the model in a larger framework (a "world picture") in a non-deductive way . . . the story should help us gain some understanding of the physical mechanisms. (344)

Hartmann points out that the relation of a story to an underlying theory "depends on the concrete example in question" (327). Jenann Ismael's (2013) notion of a *theory/user interface* is importantly similar:

> A naturalistic perspective on representational activity views representations of all kinds—from everyday concepts to the specialized products of professional science—as part of a user interface designed to act as an epistemic and practical intermediary with a partly known and locally controllable world . . . There are concepts that reflect the way things are with the world, and concepts that reflect the way things are with us, and then the great, grey area in between of structures that are simultaneously self- and world-involving in a mix that can't be described in terms that are in general any more compact than the account of their role in the coupling described above. (2013, 92–93)

We can think of all these items—a Hartmann *story*, an Ismael *theory/user interface*, and my *explanatory gloss*—as, in effect, a *user's manual* for applying a theory to its explanatory target *as that target is pre-theoretically conceived*. The nature of the theory/user interface will depend on (1) the type of *theory* and (2) the type of *user*. With respect to (1): if the explanatory target is a cognitive capacity, then an *intentional* gloss is appropriate, for the various reasons I cited in chapter 1 and about which I will say more below. A different sort of gloss will be required for fundamental theories in physics. With respect to (2), an important consideration is whether the venue for dissemination of the theory is a scholarly article aimed at a professional audience, a piece in the popular press intended for general consumption, or an introductory textbook intended for beginning students. Different needs will be served in each case; the gloss is intended to make the theory intelligible for the relevant audience.

Finally, Colin Allen (2013) explicates Hartmann's notion of a story in a way that makes explicit its relevance to our purposes:

> Scientists must not only provide interpretations of their models, which connect the mathematics to the measurable phenomena, but also stories that link the models to the questions with which their enquiries began. In cases where those questions originate with phenomena that are understood pre-scientifically or pretheoretically, a story (in this technical sense) must explain why a given model is relevant to the initial questions, even if sometimes the upshot is that these questions were based in confusion. (2013, 257)

Recall from the discussion in chapter 1 that the primary function of an intentional gloss is to illustrate, in a perspicuous and concise way, how the computational theory addresses the intentionally characterized phenomena with which inquiry began and that it is the job of the theory to explain. The explanatory targets of theories of cognition are manifest, personal-level *rational* capacities. These capacities are explained by computational neuroscience and psychology as *mathematical* capacities of sub-personal mechanisms that are in turn further explained as causal processes. The intentional gloss that accompanies the computational theory proper can be seen as "connective tissue" linking the account of the causal processes in the theory with our commonsense conception of ourselves as rational agents who are generally successful in our various cognitive endeavors but occasionally subject to error. In other words, it connects the theory with our normative self-conception. The gloss enables the theory to satisfy an intelligibility constraint specific to the explanation of cognition: in general, the outputs of cognitive processes should *make sense*, should be *rational*, given the inputs. On the deflationary account of representation offered here, that constraint can be satisfied without requiring a substantive, non-intentional and non-semantic relation that grounds the attribution of determinate content, and hence accuracy conditions, to the states and processes characterized by the theory proper. Rather, attributing pragmatically motivated distal content in the gloss allows us to see the subject's activity as based on a largely accurate representation of the world and, occasionally, sometimes systematically, as misrepresentation. Normative notions such as correctness and mistake are confined to the intentional gloss. The intentional gloss provides what we might think of as a "rational overlay" on the mathematical and mechanical processes characterized by the theory proper, an overlay that makes contact with our commonsense conception of ourselves as rational creatures, generally successful given our goals but occasionally subject to error. In doing so, it also makes contact with some of our cherished philosophical projects, for example, with normative epistemology.

One might simply deny that a theorist should concern herself with the pretheoretic conception of the explanatory target, especially if subsequent theorizing has revealed that the original conception is radically flawed. With respect to cognition in particular, if, as I have argued, the theory proper can explain our cognitive capacities without appeal to content, then why bother with content at all? Why not simply *eliminate* it, as Stich (1983) proposes, arguing that content attribution is irremediably observer relative, context sensitive, and vague, and so unfit for use in science. Noam Chomsky also denies that theorists have an obligation to make contact with the way that we see ourselves, claiming that the sciences of the mind should be purged of normative and intentional notions. Such notions as "solving a problem," "making a mistake," "misrepresenting," he claims, just reflect our parochial interests, and hence have no place in science (Chomsky 1995).

There is a sense in which eliminativists such as Stich and Chomsky are right. If our cognitive capacities are to be naturalistically explained, then such notions as *representation* and *content* will eventually drop out of the account. A naturalistic theory must discharge any debt to intentional notions. On the deflationary account, the theory (proper) has no such debt to discharge. Rather, it is our commonsense understanding that has such a debt, one that the theory shoulders, but the theory needs the intentional gloss to show that the debt has been settled.

Moreover, there is something to the idea that seeing ourselves as rational creatures who are occasionally subject to error is indeed, from a more detached perspective, *parochial*. We can't recover the distinction between rational activity and a mistake at the level of fundamental physics, nor even at the level of neural processes. Martian scientists, looking down on our explanatory practices from afar, may have no use for an intentional theory/user interface. They may be interested only in explaining and predicting our behavior considered as bodily motion. Nonetheless, *we* need the intentional interface. It puts a human face on the mathematical and mechanical processes characterized by the theory. Its job is to make clear that what has been naturalistically explained by the theory is indeed *the mind*, as understood by us. More theory won't achieve this goal. Sellars (1963), discussing the relationship between what he called the "scientific" and "manifest images," puts the point this way:

> The conceptual framework of persons is not something that needs to be reconciled with the scientific image, but rather something to be joined to it. Thus, to complete the scientific image we need to enrich it not with more ways of saying what is the case, but with the language of community and individual intentions, so that by construing the actions we intend to do and the circumstances in which we intend to do them in scientific terms, we directly relate the world as conceived by scientific theory to our purposes, and make it our world and no longer an alien appendage to the world in which we do our living. (1963, 40)

In this eloquent passage, Sellars gives voice to the particular need that an intentional gloss, as "connective tissue" bridging theory and self-conception, or Ismael's coupling of "self-and world-involving structures" is designed to address. The eliminativist reveals a kind of blindness in not recognizing that need.

2.5 Conclusion and Looking Ahead

Recall Fodor's claim that "intentionality . . . doesn't go that deep . . . if aboutness is real it must be something else" (1987, 97). The remark set the agenda for the "naturalistic project," pinpointing where in the transition from fundamental properties to sophisticated intelligence genuine intentionality enters the picture. Hutto and Myin (2020) characterize the robust representational realist's motivation for the project:

> *There must be something special* that underwrites such cognitively-driven activity; something that explains why such intelligent activity is set apart from the rest of unthinking nature that contains only mere, but not genuinely intelligent, behavior. . . .
>
> Thus the truth of mental representationalism is secured simply by the fact that we are seeking to explain intelligent behavior—namely, the competencies of intelligent creatures or systems. (2020, 86, emphasis in original)

That "something special" is, of course, *intentionality*; genuine intelligence, it seems, requires it. So, theories that explain our cognitive capacities, which are pretheoretically characterized in intentional terms, must identify the point where nature turns the trick, where representation *goes mental*, as Morgan (2014) puts it. I have argued that intentionality does not play a substantive role in the sciences that aim to provide a foundation for the study of mind. According to the picture that emerges from these sciences, there isn't a deep divide in nature between those systems that are genuinely representational and those that are not, though these sciences, when accompanied by pragmatically motivated intentional glosses, do support a reconstruction of intentionality in biological systems such as ourselves, as I explained earlier. The deflationary view locates representation in our commonsense way of understanding ourselves, and intentionality in our well-entrenched philosophical attempts to make sense of the commonsense phenomenon. I pick up this theme in the final two chapters, where I defend deflationary construals of propositional attitudes and perceptual experience.

3 Belief and Its Linguistic Representation

Beliefs and the other propositional attitudes play central roles in our commonsense practice of predicting and explaining each other's behavior. For example, we appeal to Alice's belief that vaccines are dangerous to predict and explain her refusal to get vaccinated against Covid-19. The practice works so well that we shape our lives by it. My friend Marisa and I agree to meet for lunch at a designated time and place. I make sure that I am at that place at the designated time because I know that Marisa *believes* I will be there and, like me, she *desires* and *intends* to keep her commitments.

Why does commonsense psychology (hereafter, CP) work as well as it does? Put another way: What must beliefs and the other attitudes be like for them to successfully subserve our commonsense practices? I consider two possible answers: first, an answer based on the view that construes the attitudes as relations between subjects and internal representations, according to which they literally have syntax, meaning, and truth conditions, and then a deflationary alternative that takes the attitudes to be properties of subjects that are merely *modeled* as linguistic objects, without having such linguistic properties themselves. I argue that the deflationary view provides a more compelling account of CP's success.

3.1 The *Representationalist* Answer

The *representationalist answer*—developed by Jerry Fodor fifty years ago and still the most popular view today[1]—purports to explain CP's success by claiming that beliefs and the other propositional attitudes are relations to inner representations. More specifically, according to the *Representational Theory of Mind* (hereafter *RTM*):

> RTM: To believe (desire, fear, and so on) that *p* is to bear a particular functional/computational relation to a mental representation MR that means that *p*.

For example, to believe that vaccines are dangerous is to bear the functional/computational relation characteristic of belief to a mental representation that means that *vaccines are dangerous*. To fear that vaccines are dangerous is to bear a different functional/computational relation to a token of the same MR type.

According to the RTM, it is the fixed functional architecture of the mind/brain that explains why CP works as well as it does. The explanation appeals to the following claims:

1. Our best cognitive scientific theories type-individuate mental states in the same way as commonsense psychology type-individuates propositional attitudes—that is, in terms of relations to objects with content. Importantly, the content of an attitude, as specified by the sentence that constitutes the complement of the *that-clause* in our ordinary attributions (as in, for example, Alice believes that *vaccines are dangerous*), is the same as the content of its constituent MR.[2]
2. CP is a *deep theory* of the attitudes, as Fodor (1987, 6–7) puts it. The idea has two components: (a) the theory's generalizations are defined over *unobservable* entities, in particular,

over mental representations, and interactions among them; and (b) the theory's explanations, when made explicit, have a deductive structure said to be characteristic of explanation in the sciences.

3. We *know* this theory and *use* it in social interactions to predict and explain behavior, in much the way that scientists use their theories to predict and explain observable phenomena.
4. CP is universal among humans and innate.[3]

CP, according to the representationalist picture, is "proto cognitive science." It enables the prediction and explanation of behavior because it limns the deep causal structure of the mind.

Fodor supports the RTM and argues that it explains the success of CP by appeal to what I will call the *striking parallelism* argument:

> There is a *striking parallelism* between the causal relations among mental states, on the one hand, and the semantic relations that hold among their propositional objects on the other; and . . . very deep properties of the mental—as, for example, that trains of thought are largely truth-preserving—turn on this symmetry. RTM suggests a plausible mechanism for this relation, and that is something that no previous account of mentation has been able to do. (1987, 18, emphasis added)

Familiar examples support the "striking parallelism" claim that causal relations in thought "mirror" inferential relations among their propositional objects: (1) if O believes $P \rightarrow Q$, and O believes that P, then O is likely to infer that Q; and (2) if O believes something of the form Fa, then O believes something of the form $(\exists x)Fx$. The mental states involved in these thought processes are causally related, and their contents are logically and semantically related as premises and conclusion—*modus ponens* and *existential generalization*, respectively.[4]

Representationalism proposes to explain the striking parallelism—the fact that causal relations among mental states in thought processes "mirror" logical relations in inferences—by positing that thought processes literally *are* inferences. They are largely truth-preserving because, as in first-order logic, they are sequences of symbols whose transitions are governed by rules that are defined over the *syntax* of their constituent mental representations, where syntax is understood as a type of higher-order physical property. The idea that MRs have syntactic properties, essential to the representationalist answer, is a further assumption that goes beyond the RTM itself. Beliefs, on this view, have the characteristic properties of linguistic representations: in particular—syntax, meaning, and truth conditions. The RTM is claimed to *vindicate* CP (Fodor 1987, ch. 1), where vindicating involves explaining why it works. Central to the RTM's explanation of that success is that beliefs (and the other attitudes) have linguistic properties.

3.2 A *Nonrepresentationalist* Alternative

An alternative, nonrepresentationalist, explanation of why CP works as well as it does has the following components, which I will explain and motivate in turn:

1. CP is not a deep theory of the attitudes; rather, it is theoretically *shallow*.[5]
2. CP plays a regulative, *mind-shaping* role in our commonsense practices, in addition to its roles in prediction and explanation.
3. CP is, at least in part, a product of enculturation.

Turning first to (1). A deep theory of some entity or property characterizes it in terms of a hidden nature. Thus, chemistry characterizes water in terms of its chemical composition

(H_2O) rather than its typical observable manifestations. Shallow properties are properties whose possession conditions are closed tied to observable phenomena. Beliefs and the other attitudes, as they figure in our commonsense psychological practices, are properties that play characteristic roles in the causation of behavior. On this non-representationalist alternative, they are closely connected to their typical behavioral manifestations, rather than relations to unobservable mental representations. They are shallow rather than deep states.

There are compelling reasons for denying that CP is a deep theory. Despite centuries of successful CP practice, we are not now and never have been in a position to know how the attitudes are realized in cognitive architecture. For all we know given the current evidence, just as there are various ways that information may be realized in machine architecture, so there may be other ways that information may be realized in human cognitive architecture, perhaps as a characteristic pattern of activation of a widely distributed neural network. The possibility that some believers have a cognitive architecture distinct from that presumed by the RTM cannot be ruled out a priori. Maybe you or I have such an architecture. If that were the case, what would follow? A proponent of the RTM would seem to be committed to denying that such a subject has beliefs. In the absence of strong empirical evidence that believers *do* instantiate the RTM architecture, the RTM, far from vindicating CP, invites skepticism about the attitudes. Given the very real possibility that our mental processes are subserved by quite different architectures, we would have no basis for confidence that our apparently well-supported attributions—those with overwhelming behavioral evidence—are in fact true. All we could be confident of is that they are useful for predicting behavior.

This is surely a mistake. CP is not hostage to whatever cognitive science eventually discovers about our cognitive architecture. Whatever our cognitive architecture, CP would not be imperiled because it is not committed to any particular view of cognitive architecture. So, CP should not be construed, as Fodor would have it, as *proto cognitive science*. The "folk" don't have a view about our cognitive architecture—they don't have a theory, much less a "deep theory," of how the attitudes are realized in our brains. More importantly, they don't care; such matters are not germane to the practical roles that the attitudes play in CP. There is an important disanalogy between belief and water. The chemical composition of water *is* germane to the roles that it plays in our lives, and its hidden nature was thought to be important even before its chemical composition was discovered. It was assumed that water's hidden nature, whatever it is, determines its boiling and freezing points, the substances that it can dissolve, and so on, matters of great practical concern. There is no reason to think that an analogous point is true of beliefs. CP entails no commitments as to their nature, beyond their being causally efficacious, in conjunction with desires and other mental states, in the production of behavior. CP is not committed to their being internal representations of some sort. Beliefs, as they figure in our commonsense predictive and explanatory practices, are not deep states of the sort presumed by the RTM.

Here is an explicit argument for the claim that the commonsense notion of belief is shallow rather than deep:

1. Given the central role that beliefs play in our commonsense practices, they must be easily and reliably attributable on the basis of the available evidence.
2. Attributors typically have observable evidence of behavior and often a raft of experience about what they themselves and others would think and do in similar circumstances.

3. This evidence is often sufficient to justify belief attributions to others.
4. But belief attributors *never* have the sort of evidence—evidence about underlying cognitive architecture—that a deep theory like the RTM would require for attributions of belief to be justified.
5. Therefore, CP is most plausibly committed to a shallow notion of belief and the other attitudes; it is not committed to the attitudes having a particular hidden nature.

The yawning gap between what deep-state theorists claim is attributed in our commonsense practices—that is, states with a particular hidden nature—and the sort of evidence attributors typically have for belief attributions would make such attributions highly speculative. Rather than explaining the remarkable success of CP, a deep-state construal would (and should) give rise to skepticism about whether there really are beliefs—that is, whether there really are states of the specified sort. At best, it would support only an instrumentalist or fictionalist interpretation of CP: it is useful for predicting behavior but should not be construed as saying anything about the causes of behavior.

A shallow construal of belief conforms to a reasonable conservative methodological principle: interpreters of CP should assume that it attributes no more structure to beliefs (and the other attitudes) than is required for them to play their characteristic roles in our belief-attributing practices. To play their characteristic roles in our CP practices, beliefs and the other attitudes must be construed as causally efficacious properties of subjects that are manifest in observable behavior. There are several ways that these properties may be understood.

Beliefs may be construed as *complex functional properties*; for example, the belief that *it is raining* is the property whose instantiations are typically caused by feeling rain on one's skin when out walking, or seeing that the streets are wet, or being told that it is raining, and that typically causes (in conjunction with the desire to avoid getting wet) one to grab an umbrella, to say that it is raining, to believe that one will get wet if one stays outside, and so on. This view is sometimes called *role functionalism* as opposed to *realizer* (or *filler*) *functionalism*.[6]

Alternatively, beliefs may be construed as *complex dispositions to behavior and to other mental states such as thought and affect*.[7] According to this view, the belief that there is beer in the fridge is the property whose instantiations are constituted by dispositions to go to the fridge when one wants a beer, to say "Would you like a beer?" to a visiting friend, to be surprised to discover, when one goes to the fridge for a beer, that there is none there, and so on. The essential difference between dispositional properties and the functional properties typically associated with role functionalism is that dispositional properties are forward-looking, focusing on downstream effects on behavior, thought, and affect, while the relevant functional properties are both forward and backward looking, including also upstream causes, for example, seeing or being told that there is beer in the fridge. That said, I see no reason why a dispositionalist could not include, among the dispositions constitutive of an attitude-type, dispositions to be caused by particular antecedent conditions—for example, by seeing beer in the fridge, or, in the case of believing that it is raining, by seeing that it is raining. On this view, the distinction between functionalist and dispositionalist accounts of the attitudes collapses.

Whether beliefs are construed as functional roles or as complex dispositions, the constituent elements of a particular belief type are just the causal roles or dispositions that competent users of belief attributions would take to be constitutive of that belief type. In general, beliefs and the other *cognitive* attitudes are responsive to states of the world, and when things go well carry information about the world,[8] while desires and the other *conative* attitudes are, *ceteris paribus*, action-producing and directed at changing the world.

I will remain neutral on whether the attitudes are best construed as functional roles or as complex dispositions, referring to them in what follows as *causal complexes*, a term intended to subsume both. Importantly, the complexes—whether they are construed as functional roles or as dispositions—are shallow; they include causal relations to the sorts of things recognized by commonsense and characterized, in part, by their relations to observable behavior but not any relations to mental representations or other unobservable structures or states that may be posited in scientific accounts of cognitive processing.

Turning to the second component of the non-representationalist alternative:

2. *CP plays a regulative, mind-shaping role in our commonsense practices.*

Victoria McGeer (2007; 2021) and Tad Zawidzki (2008) have argued that the attribution of belief and the other propositional attitudes serves the important normative function of *regulating* behavior. Indeed, they argue, regulation of behavior, rather than prediction and explanation, is the primary function of commonsense psychology. As McGeer puts it:

> We overlook the way folk psychology operates as a regulative practice, molding the way individuals act, think and operate so that they become well-behaved folk-psychological agents: agents that can be well-predicted and explained using both the concepts and the rationalizing narrative structures of folk psychology. (McGeer 2007, 139)

There are two distinct aspects to this regulation: (1) Commonsense psychologists—that is, all of us who employ CP for predicting and explaining behavior—have fairly well-developed normative views about what others *ought to do, think, and feel* in various circumstances, and these views drive their judgments about particular cases. So, for example, if a subject believes that if P then Q, and believes that P, then we commonsense psychologists will hold that this subject *should* (*ceteris paribus*) believe that Q. (2) Subjects (including oneself) generally *do* behave, think, and feel as they should (in our example, believing Q) and so assuming conformity to these norms works well for predicting and explaining their behavior. And so, for example, I can safely presume that my friend Marisa will show up for our lunch appointment, and I conduct myself accordingly.[9]

McGeer considers how we react when behavior violates the norms. In support of her claim that CP is primarily a regulative practice, rather than a "proto-scientific" enterprise aimed simply at prediction and explanation, she points out that when the normative principles fail to predict a subject's behavior we do not attempt to revise the principles, as a theoretician might do when her theory's principles fail to predict the phenomena. Rather, we often try to mold the subject's behavior to fit the principles, deploying various pressures, primarily criticism, but also—especially for children—cajoling and encouraging, to increase the chance of conformity in the future. CP is largely, as Zawidzki (2008) puts it, a *mind shaping* enterprise. In part as a result of these pressures, we become good models of commonsense psychology. The norms come to function as causal generalizations that enable the reliable prediction and explanation of behavior.[10]

Summarizing how the mind-shaping process works: We assume conformity, expecting others to be readable and predictable. We endeavor to conform ourselves; that is, we make ourselves readable and predictable. As a result, we become good models of CP. In virtue of our being such models, the norms come to structure our interactions with each other without us thinking about it. This is why CP works as well as it does.[11]

Turning to the final component of the alternative picture:

3. *CP is, in part, a product of enculturation.*

We can wonder where our facility with CP—both in applying the norms to others and in conforming to them ourselves—comes from. Proponents of RTM, recall, claim that it is universal among humans and innate, the taxonomic scheme of beliefs, desires, and the other attitudes reflecting the fixed functional architecture of the mind. The view that CP involves mind-shaping implies that enculturation plays a role, but a range of views is possible here.[12] *Social constructivists* claim that it is primarily a product of enculturation (see Vygotsky 1978; Gergen 1985; 1997). McGeer (2007) and Matthews (2013) endorse a mixed view, involving some interplay of innately determined capacities and enculturation. Here is McGeer:

> Human infants do not confront a world of "unstructured experience", and not just because they have innate mechanisms for ordering whatever experience is given to them. Their own ordering capacities are given a significant boost, not just once but again and again over the course of development, by parents who shape their children's experience by involving them in structured interactions governed by the sense-making norms of folk psychology. (McGeer 2007, 153)

It should be noted that the non-representationalist construal of CP articulated here does not require denying that there is a substantial innate endowment; it requires only that whatever innate endowment there *is* is significantly boosted by enculturation. One needs only to look to our own culture to see that, with respect to both adults and children, significant social pressure is brought to bear to make it more likely that we are good models of the CP framework. This is something that even proponents of the RTM should accept, whether we call it mind-shaping or not.

Matthews, who is a dispositionalist about the attitudes, considers how dispositions cluster together in characteristic ways to constitute the various attitudes. For example, if one is disposed to say "there is beer in the fridge," then one is also disposed both to go to the fridge if one wants a beer and to be surprised if one does so and finds no beer there. He explains this as follows:

> What we come to have when we come to be reasonably good models of our commonsense propositional attitude psychology is a propensity to package these dispositions to behavior, thought, and feelings in certain fairly predictable ways, ways that facilitate our social lives and the social lives of others. There are no doubt different ways that different cultures might (and indeed do) package these dispositions, if they attend to dispositions at all—ways that would lead to different commonsense psychologies, some of which might not traffic in propositional attitudes at all. (2013, 117)

The upshot is that in becoming good models of CP, subjects acquire both (1) the norms that govern their own behavior, shaping their behavioral and mental dispositions, and moreover, enabling the prediction and explanation of others' behavior, and (2) the scheme whereby elements are packaged together to form the causal complexes—the belief that p (e.g., that there is beer in the fridge, that it is raining), the desire that q (e.g., that one acquire a beer, that one not get wet)—to which the norms apply.

Putting the various pieces together, here is how the non-representationalist alternative explains the success of CP. Beliefs and the other attitudes are not "deep states" but shallow states, characterized by their connections to observable behavior—what I am calling *causal complexes*. In general, they are reliably attributable on the basis of the available evidence, which is primarily behavioral. We are *shaped*, partly as a result of enculturation, to be good models of CP; hence the generalizations of CP serve to predict and explain our behavior reasonably well.

Let's compare the two explanations of CP's success. The RTM promises a scientific vindication of our commonsense practice, appealing to speculation about our cognitive architecture that goes well beyond the current scientific evidence, and certainly, well beyond any evidence available to CP's practitioners. The more minimal, non-representational alternative appeals only to observable features of the commonsense framework itself. The RTM's explanation *requires* that propositional attitudes have linguistic properties; the more minimal alternative does not, leaving future science to discover how the attitudes are realized in human cognitive architecture.

In the rest of the chapter, I explore the possibility that content might be construed as an *intentional gloss* on the causal complexes that, according to the account developed here, constitute the attitudes. In other words, I sketch a *deflationary* account of belief.

3.3 Content as *Type-Specifying Gloss*

To play their roles in our commonsense practices, beliefs and the other attitudes must be properties of subjects that cause behavior and other mental states, including emotions, moods, and so on. It is common ground that the causal relations that hold among the attitudes are modeled by and represented as logical and semantic relations that hold among the sentences used to attribute propositional attitudes.[13] This is a minimal commitment of Fodor's *striking parallelism*. The parallelism that Fodor notes allows us to represent a subject's mental processes by *modus ponens, existential generalization*, and other inference schemas, whether these causal transitions really are inferences defined over internal sentences or not. The representationalist explanation of the striking parallelism holds that causal transitions in thought are largely truth-preserving because they are governed by rules that are defined over the syntax of the internal sentences relations to which constitute the attitudes. Beliefs, representationalists hold, have the characteristic properties of linguistic representations, in particular, syntax, meaning, and truth conditions. But the alternative picture can explain the striking parallelism too, without appeal to speculation about our cognitive architecture: we are *shaped*, partly as a result of enculturation, to conform to CP's generalizations, which are typically characterized in terms of logical relations that hold among the sentences used to attribute the attitudes.

Once the plausibility of representationalism—as an account of the attitudes as attributed in our commonsense practices—has been undercut, and we replace it with a more minimal account of what commonsense is committed to, then the motivation for thinking that beliefs have syntax, content, and truth conditions is also undercut. A less committed position becomes a possibility. We can think of beliefs as properties of subjects that are *merely* modeled by and represented as objects with these linguistic properties, without their actually having these properties themselves. In other words, the more minimal account of CP's commitments supports a *deflationary* conception of belief. It is deflationary in construing our linguistic representation of belief as a kind of pragmatically motivated *gloss* on the causal or dispositional properties of subjects that play essential roles in the prediction, explanation, and regulation of behavior.

The view that content is a gloss on causally efficacious properties of thinking subjects shares its fundamental motivation with the *representational theory of measurement*, a view that holds that relations among quantities of physical magnitudes can be represented in numerical terms. In general, such a scheme works because there is a real empirical structure of relations in the world that is structurally similar to the numerical scheme (typically, relations defined on the real numbers) used to represent it. In the domain of commonsense psychology, this is the much-noted *striking parallelism* between causal transitions in thought and logical relations among attributions. Here, the empirical structure of relations among causal or dispositional properties of CP subjects arises from the effects of mind-shaping on whatever innate endowment such subjects possess. In both cases, there is a real empirical structure (of relations) that explains why the scheme is useful. Becoming a good model of CP involves shaping one's causal complexes, however they are characterized, so that the relations among them conform to the logical relations that hold among the linguistic representations of these properties, in much the same way that numerical relations hold among numerical representations of quantities of physical magnitudes.

On this deflationary view, the complement clauses of attitude attributions (*that-clauses*), specifying the contents of the attitudes, function in our commonsense attributive practices much like the real numbers function in our measurement practices.[14] In our everyday measurement practices we specify physical magnitudes such as mass, temperature, and length in ways that allow us to capture such relations as *heavier than*, *warmer than*, and *longer than*. We do this by representing particular quantities of magnitudes on the real numbers. Importantly, we can represent these magnitudes knowing little or nothing about their intrinsic character. The commonsense practice of representing mass and temperature on the real numbers long predates any scientific theory of what mass and temperature are.

Let me spell out the numerical analogy in more detail. To say that an object has a temperature of 20°C is not to say that the object has a certain numerical property. Rather, it is to attribute to the object a property that is specified by means of its location on a numerical scale, allowing us to think, talk, and reason surrogatively about objects with temperatures without having a "deep theory" of temperature.[15] We gloss the physical magnitude temperature using the real numbers, but we don't assume that the magnitude (that is, temperature itself) has all the properties and relations of the representing scheme—the reals—that we use to think about it. We can infer that an object with a temperature of 20°C is warmer than an object of 10°C and colder than an object of 40°C, but we are not licensed to conclude that it has twice as much heat/temperature as the former object or half as much as the latter. Determining which properties of the representational scheme we are licensed to attribute to the empirical domain requires empirical investigation and argument.

Analogous points can be made about our commonsense practice of specifying beliefs in terms of *that-clauses*: Just as in the measurement case, where the numerical specification doesn't pick out a numerical property of the object, similarly in the belief case, the *that-clause* serves to specify the belief but without justifying the conclusion that the belief has the linguistic properties of the *that-clause* used to specify it. To say that a subject believes *that it is raining* is not to say that she stands in some substantive relation to something that means *it is raining*. Rather, it is to say that the subject possesses a causal complex that is specified in terms of the *that-clause*. We represent the attributed attitude—we *gloss* it, in my terminology—in linguistic terms, though without commitment to the idea that the attributed attitudes really have these linguistic properties. In so doing, we provide a way of representing the attitudes for our practice of predicting, explaining, and regulating behavior.

On this deflationary view, causal relations among the attitudes are represented in terms of logical relations among the *that-clauses* that we use to specify the attitudes. I have argued that the relevant causal relations are shallow, closely tied to behavioral manifestations, not to be construed as relations holding among "deep states" such as internal sentences. On this view, beliefs themselves don't have content and truth conditions. And beliefs don't stand in logical relations to each other. Rather, the elements in our scheme for representing them do. Robert Stalnaker seems to have something like this idea in mind when he says:

> There is, alas, only a brain in the head—no pieces of information, no contents, wide or narrow. Propositions, whether they are Fregean thoughts, Russellian complexes, or sets of possible states of the world, are abstract objects that we use to represent certain capacities and dispositions of people, and certain kinds of relations between them. (2008, 109)

To see why *that-clauses* can go proxy for a complex description of causal complexes, what Stalnaker says here must be augmented with the point stressed by McGeer and Matthews. The linguistic strategy for representing the attitudes works as well as it does precisely because we can be shaped to conform our dispositions to the logical relations holding among attitude attributions—that is, to conform to CP's generalizations, packaging our dispositions in the right way. *That-clauses* serve to represent causal complexes which, through some combination of natural inclination and enculturation, tend to co-occur in well-regulated subjects. For example, Bill believes it is raining just in case he is disposed to say that it is raining if asked about the weather, to believe he will get wet if he stays outside, to grab an umbrella if he desires to stay dry, and so on. Belief *attributions* will be straightforwardly true or false depending on whether the subject possesses the causal complex so designated.

Content, on the deflationary view defended here, does what we need content to do, but without the heavy metaphysical and empirical commitments of representationalism. Content attributions serve at least three related functions:

1. Most importantly, *content attributions provide a compact way of picking out the causal complexes that constitute the attitudes and that play crucial roles in the prediction, explanation, and regulation of behavior.*

The property whose instantiations are typically caused by looking out the window on a rainy day, or by listening to a weather report of rain, or by feeling moisture on one's skin while out walking, and that typically causes (in conjunction with certain other attitudes) umbrella-carrying behavior, reporting that it is raining, and so on, can be referred to as the belief *that it is raining*. I have identified only a fraction of the belief's causal role, but the content specification allows us to infer quite a bit about how the belief will interact with environmental conditions and other mental states to produce additional mental states and behavior. We don't have any other access to, or way of characterizing, the causal complexes in virtue of which beliefs serve their predictive, explanatory, and regulatory functions. Certainly, current science is not in a position to specify what is going on in their brains when subjects believe it is raining.[16] And for all we know, the mental architecture underlying the relevant complex may differ from one subject to the next. In any case, as commonsense psychologists, we simply don't care about the architecture underlying the causal complexes that constitute the attitudes; knowledge of deep causal structure would not further our predictive, explanatory, and regulatory ends.

A couple of caveats are in order. The complement clause of the attribution—the *that-clause*—picks out the causal complex only imperfectly. Subjects may not instantiate all the elements of the package characteristic of a particular belief. A subject to whom it is appropriate to attribute the belief that it is raining may not be disposed to carry an umbrella. She may prefer the chance of getting wet to the inconvenience of carrying one. And further observation may reveal more of the relevant cluster, leading us to refine the attribution. Initially inclined to attribute to a subject the belief that there is beer in the fridge, we have good grounds for attributing the more specific belief that *there is Stella Artois in the fridge*, after hearing him offer a "Stella" to a visiting friend.

2. *Content attributions point us to the relevant part of the world that the attitude is "directed at."*

This second function will take some spelling out. It is a truism that beliefs are about the world. The content-attributing *that-clause* of a belief points us to the relevant part of the world that the belief is directed at. The belief *that there is beer in the fridge* is directed at the fridge and some of its contents. The causal complex constitutive of the belief that there is beer in the fridge involves dispositions to act, think, and feel in particular ways with respect to the fridge and some of its contents.

On the deflationary view, beliefs themselves do not have intentionality or aboutness, although there is generally no harm in thinking of the meaning of the sentences that attribute beliefs as conferring a kind of intentionality on beliefs themselves, just as we may come to think of physical magnitudes as having numerical properties.[17] One practical payoff of this worldly directedness of belief, as displayed by the belief-specifying *that-clause,* is that it helps us to identify the relevant causal complex that type-identifies the belief—in other words, it serves the first function of content attribution, described earlier—and it does so in terms of the complex's publicly available causes and effects. So, this second function of content attribution—specifying the aspects of the world that the belief is directed at—and the first function are intimately connected. But this second function, concerning the directedness of belief, is distinct from content's function of enabling us to identify the causal complex underlying behavior. Believers do not exist in isolation from the objects and properties that make up our shared world. The content attribution situates the subject in that world. The suite of attitudes attributable to someone playing soccer is directed at various objects around her: the ball, other players on the pitch, if she is an offensive player then the opponent's goal, if she is playing defense then also her own goal, and so on.

A perhaps obvious objection is that this "directedness" of belief is, in fact, *intentionality*. And, indeed, this would be a natural way to characterize what I am calling "directedness," if beliefs and the other attitudes had content and truth conditions. On the deflationary view I am promoting, they do not. There are certainly relations between states of subjects and worldly conditions that are not intentional. For example, hunger is directed at food, without *meaning food*. But once we have a linguistic representation of the attitudes, in terms of content attributing *that-clauses*, it tends to sweep up a whole host of dispositions (or more generally, causal roles) that subjects have toward the objects in question, and we tend to think of the relation as an intentional relation.[18]

It is this directedness of the subject toward aspects of the world that explains in part how the individual causal roles that make up the causal complexes constitutive of the attitudes are packaged together in the ways that they are, and to which the norms of CP apply. A subject who believes that there is beer in the fridge may also believe that there is beer at the local bar. Though there may be some overlap in the causal complexes constitutive of the two beliefs, they will still be quite distinct. The subject may also believe that New Brunswick is thirty-five miles from New York City, and so instantiate a suite of dispositions (causal roles) directed at New Brunswick and New York City, none of which are constitutive of her attitudes toward beer.

There are, of course, cases where subjects have attitudes toward nonexistent objects. A naïve subject may believe that Superman works at *The Daily Planet*, not realizing that Superman and *The Daily Planet* are fictional objects. Such cases pose no special problem for the deflationary view. In this case, there are no objects for the content attribution to point or direct us toward. The attribution still serves the first function of picking out the relevant causal complex underlying the subject's behavior. She will be disposed to respond in predictable ways when asked about Superman and *The Daily Planet*, to think about the sort of work that Superman does in his day job if the subject comes up, and so on, where these additional attitudes are to be cashed out in the way appropriate to the deflationary account.

If beliefs don't really have content and truth conditions, what is it to say that Alice's belief *that vaccination prevents Covid infection* is true? Strictly speaking, it is to say that the sentence that best models or glosses Alice's belief, the complement clause of the attribution—in this case, "vaccination prevents Covid infection"—is true. In general, to say that a belief is true (or false) is to say that it—that is, the causal complex that constitutes the belief—is best modeled, represented, or *glossed* by a sentence or proposition that is true (or false). We can say, therefore, that *what is believed* is true (or false), thus preserving the way that we normally talk about beliefs.

In support of this idea, let me call attention to a kind of ambiguity in our talk about beliefs, between (1) a causally efficacious property of a subject—that is, a propositional attitude—and (2) *what* is believed—that is, a *proposition*, as expressed by a meaningful sentence. (2) is abstract, and we use it to talk and think about (1). The distinction in language is subtle—we use the same words ("belief," "thought") to talk about both. In fact, non-philosophers don't often talk of "believing." They typically say such things as "She *thinks* that vaccines are dangerous."[19] We can disambiguate by talking explicitly about *what is believed* (i.e., a proposition), *what is desired* (a state of affairs), *what is feared* (a state of affairs), and so on. These are abstract objects, as Stalnaker noted, that we use to represent certain capacities and dispositions of people, and as the measurement analogy makes clear, they can do that job without the subject (or her brain) bearing any substantive relation to these abstract objects.

3. *Representing the attitudes by objects with meaning and truth conditions—by that-clauses—allows us to treat them as having correctness conditions, and to evaluate causal transitions among them as rational or irrational.*

The attitudes themselves are not semantically evaluable, and causal transitions among them are not literally rational (or irrational), except as one is mind-shaped to behave, think, and feel in rational ways. The content characterization provides a normative "overlay" on the causal complexes that constitute the attitudes, thus "outfitting" them for evaluation; they can then be endorsed or criticized.[20] This is necessary for CP to play its regulative or mind-shaping role. Recall McGeer's point that as commonsense psychologists we do not consider revising a generalization when it fails to predict a subject's behavior, rather we often double down on it, insisting that the subject's behavior should respect it, and criticizing the subject when it does not. In response to such regulative pressure, subjects tend, we hope, to become better exemplars of CP, and its normative schemas become increasingly descriptive, better subserving prediction and explanation. As I've said, they become *good models* of CP.

The commonsense scheme works as well as it does because, through natural inclination and enculturation, people come to package their dispositions in ways specified by the scheme. We are the kind of creatures that can be so shaped. But occasionally the scheme fails. A subject may have some of the elements of the causal complex constitutive of a particular belief-type though key elements of the complex are missing, leaving us uncertain just what to say. Monothematic delusions are striking examples of this phenomenon. Subjects suffering from Capgras syndrome insist that a loved one has been replaced by an alien, yet they often have few of the dispositions one would expect of someone with such a belief, typically failing to report the situation to the authorities and continuing to cohabit with the alleged imposter. Subjects suffering from Cotard's syndrome claim that they are dead, yet they continue to behave in ways conducive to promoting well-being (eating, staying out of the way of moving vehicles, and so on), exhibiting few of the dispositions that we would expect of someone with such a bizarre belief.

CP's predictive and explanatory schemas also fail for so-called *implicit bias* cases. A subject may claim, possibly quite sincerely, to believe that women and minorities are as capable as white men, even as their behavior belies the claim. The subject appears to have some of the elements of the causal complex constitutive of the belief *that women are as capable as men of holding positions of authority* but key elements, such as the disposition to behave in ways that promote women's access to positions of authority, are missing. It is unclear in such cases what the subject believes.[21] The regulatory dimension of CP (and of social norms more generally) explains why subjects, once they become aware of the tension between their explicitly expressed views and their behavior, typically feel pressure to achieve consistency not by becoming explicit sexists or racists but by trying to change their behavior, resolving the inconsistency in ways that conform to social and moral norms.

More generally, monothematic delusions and implicit bias have received quite a bit of discussion in the recent literature on belief precisely because, in such cases, our commonsense scheme for attributing belief, and hence for predicting and explaining behavior, systematically breaks down. In general, the efficacy of the commonsense scheme depends on people being reasonably good models of the scheme.

The deflationary view that it is only the linguistic representatives of the attitudes—their images in our model for thinking and talking about them—and not the attitudes themselves that have semantic and inferential properties is nonetheless *realist* about the attitudes themselves. It is not a version of *interpretationism*, a view sometimes associated with Dennett and Davidson that construes attitude attribution as simply a useful device for making sense of behavior.[22] The attitudes themselves—the causal complexes that figure in the prediction, explanation, and regulation of behavior—are real properties of subjects, even though the particular ways that elements of the complex are aggregated to constitute the attitude-types do reflect our pragmatic interests in making sense of each other. Beyond characterizing their overall shape, CP doesn't tell us very much about these properties. As I have argued, CP is a shallow framework, serving a variety of practical purposes. Disposition-talk and functional role-talk seem to characterize the causes of behavior, thought, and affect at the right level of description for our purposes, relating them to (actual and possible) observable manifestations. How the attitudes are realized in fixed cognitive architecture is of no practical concern.

3.4 An Objection: The Deflationary View Is Not Empirically Adequate

While it may be granted that appeal to internal sentences goes beyond anything that commonsense is committed to, it is sometimes claimed that evidence from cognitive science establishes that beliefs (and the other attitudes) are in fact relations to internal sentences, and so the deflationary view that the attitudes are *merely* modeled by linguistic properties is not consistent with current science. For example, Quilty-Dunn and Mandelbaum (2018) argue that "the empirical facts seem to demand a theory of belief that is both deep and representationalist." (2360) Some of their arguments are straightforwardly question-begging, citing as "evidence" that beliefs have characteristic properties of linguistic objects—for example, *opacity* and *truth-evaluability* (2363–2364)—and then arguing that these manifest properties of belief can only be explained on the assumption that beliefs are relations to internal representations, as the RTM holds. Of course, beliefs are represented and attributed by that-clauses, which create opaque contexts and are truth-evaluable, but beliefs themselves needn't have these linguistic properties. This is precisely what is at issue in the present context.

That said, the deflationary view is compatible with cognitive science discovering that beliefs and the other attitudes are in fact *realized* in the brain as relations to internal sentences.[23] But it is also, I have been arguing, compatible with empirical science discovering that beliefs and the other attitudes are realized in ways not dreamt of by either common sense or philosophy.

Some of the goals that Quilty-Dunn and Mandelbaum expect a theory of belief to accomplish could only be done by a deep theory, in particular, by a theory of cognitive architecture and psychological processes. For example, they expect a theory of belief to explain how beliefs are stored (2358), and to explain how—that is, in virtue of what causal-mechanical processes—beliefs and the other attitudes interact to cause behavior (2361–2362). But there is no reason to think that these explanatory tasks are within the job description of a theory of belief. Belief attribution finds its home in commonsense psychology; that is where we should look to see what appeal to belief is supposed to do. What we find is that it plays crucial roles in the prediction and explanation of behavior and in regulating behavior, thought, and affect. Not only is a deep theory not required for satisfying these goals but also it would be an impediment. As I have argued, to do these jobs well, beliefs must be construed as shallow states, tied fairly closely to observable behavior.

3.5 Further Issues

3.5.1 Linguistic Meaning

The account of the attitudes defended here takes the meaning of attitude *attributions* for granted. A popular strategy for explicating linguistic meaning is Grice's proposal: explain linguistic meaning in terms of the beliefs and intentions of speakers. The deflationary account might be thought to preclude the Gricean strategy, since propositional attitudes don't themselves have semantic properties, or have them only by courtesy, in virtue of the meaning of the *that-clauses* that serve to pick out the causal complexes that constitute the attitudes. Explaining the meaning of attitude attributions, and linguistic meaning more generally, by reference to the attitudes themselves may appear, on the deflationary view, to be blatantly circular.

In fact, the circularity is only apparent. Grice proposed explaining linguistic meaning by appeal to the propositional attitudes of speakers, as we use and attribute them in our ordinary practices. He had no independent views about the metaphysics of the attitudes. The deflationary view does not preclude explaining linguistic meaning by reference to the beliefs and intentions of speakers. On the deflationary view, propositional attitudes, including beliefs and intentions, are causal complexes that don't literally have content, although we access and talk about them using meaningful attitude attributions. In other words, we *use* the linguistic content of attitude attributions to specify functional properties of speakers—properties that play crucial roles in the prediction, explanation, and regulation of behavior and other mental states—properties that are not themselves semantic or contentful, analogous to the way that we use the real numbers to specify temperatures, masses, and other properties of physical systems that are not themselves numerical. The Gricean, who seeks to explicate linguistic meaning by reference to the beliefs and intentions of speakers, is free to do so on the deflationary view, since these are just properties of subjects individuated essentially in terms of their functional or dispositional roles. There is no circularity.

We can give a similar account of other public signs. A hand movement might be interpreted as a greeting or as a threat. To interpret it as a threat is to attribute to the subject a complex functional property involving hostile affect and dispositions to further intimidating behavior. In general, to interpret a bodily movement as a gesture is to construe it as caused by particular intentions and beliefs (and affective states) of the subject, where these are functional properties that do not have content or meaning themselves, though they are specified by means of attitude attributions that do.[24]

3.5.2 Animal Belief

Animal belief is a large and fascinating topic, beyond the scope of the present chapter. Here I just scratch the surface, noting how the deflationary view of the attitudes would apply to nonhuman animals.

Fido wags his tail when his owner reaches for the leash because, we say, he believes he is about to be taken for a walk. We attribute beliefs and other attitudes to nonhuman animals for the same reason that we attribute attitudes to each other—to predict and explain their behavior.[25] The deflationary view of the attitudes makes better sense of our attributive practices to animals than does representationalism. When we attribute an attitude to a subject, whether human or animal, we attribute a property responsible for the subject's behavior, what I have been calling a *causal complex*, rather than a substantive relation between the subject and an internal sentence or a proposition. As with attributions to human subjects, the complement clause of the attribution—the *that-clause*—picks out this property only imperfectly. One obvious difference is that for animals the complex will never include dispositions to assent to the sentence we use to characterize the property. But this is a difference of degree, not of kind; a similar point holds for prelinguistic humans and for humans who don't speak the language of the attributor. As with all subjects, further observation of animals' behavior may reveal more of the causal complex, sometimes leading us to revise and refine the initial attribution. Attributions of beliefs to animals are very often good enough to allow us to predict and (seemingly) to explain their behavior.

Davidson (1982) (following a discussion in Malcolm 1973) considers an example of a dog chasing a squirrel up a tree. We may be reluctant to attribute to the dog the belief that there is a *squirrel* up the tree. Do dogs really have beliefs specific to the species *squirrel*, as opposed to some other way of thinking about the situation, perhaps a way not expressible in our language? It seems inappropriate to attribute *our* concept to the dog. What follows is only that we are unable to *specify* the content of the dog's belief, since we only have *our* language to do so.[26] On the view that content is a gloss on the causal complexes underlying behavior, this is not particularly troubling. We are unsure what content attribution best captures the causal complex underlying the dog's behavior. To be justified in attributing a belief to the dog, we needn't assume that the belief has some determinate content, perhaps not expressible in our language, that would capture that complex perfectly. It is no different for people. Some of our attitudes may be so rich and complex, or underspecified in various respects, as in the monothematic delusions and implicit attitude cases discussed previously, that they also elude precise specification by content-attributing *that-clauses*. Content attributions, the only practical way we have of specifying the causes and effects of behavior, are an imperfect way of characterizing these causes and effects. An implication of the view that content is a gloss on the causal complexes that constitute the attitudes is that content may sometimes be *indeterminate* in various degrees, with equally good alternative ways of characterizing the causal complex, using distinct content-attributing *that-clauses*.[27,28]

3.5.3 Narrow vs Wide Content

An important question is what kind of content best glosses the causal complexes that constitute the attitudes. This is a big topic, but let me conclude with one consideration that I think points to a significant advantage of the deflationary view. *Externalists* about mental content argue that the content of thought is determined in part by reference to features of a subject's external context, by her physical environment or the community with which she shares a language. Externalists hold that physically identical subjects in different environments may have thoughts with different ("wide") contents. Earth-bound Oscar has *water* thoughts, while his counterpart on Twin-earth, where the local oceans and rivers are filled not with H_20 but a superficially similar but chemically different liquid, has *twater* thoughts.[29] *Internalists* insist, at the very least, that the content that individuates a subject's thoughts in predictions and explanations of her behavior supervenes on her intrinsic properties—that the content of thought is *narrow*—and that physically identical subjects share thought contents.

The deflationary view enables an elegant resolution of the wide vs narrow content dispute. Recall that, on this view, contents serve to pick out the causal complexes that constitute the attitudes, but content is not to be construed as an essential property of the attitudes themselves. Wide (externally determined) content *glosses* on the causal complexes underlying behavior allow us to characterize these states in terms of their publicly available causes and effects, while at the same time respecting the intuition that has motivated internalists to postulate narrow content—namely, that physically identical "twins" in different environments are psychologically identical. Different wide contents (*water* vs *twater* contents) can serve to pick out the same underlying causal complex in the twins, much as different numbers—say 212 and 100 in the Fahrenheit and Celsius scales, respectively—can be used to pick out the same underlying physical magnitude. There is no need to postulate narrow content to play this important role. We *could* represent the causal complexes in narrow terms—there may be all sorts of ways of representing the complexes—but a whole host of further social purposes are best served by characterizing the complexes in ways congruent with the ways that we conceptualize the objects and properties that make up our shared world. Nothing prevents assigning narrow contents to beliefs if some useful purpose is served by doing so. Content, on the deflationary view of belief, is always pragmatically motivated. The view allows the central intuitions motivating both externalism and internalism to be jointly satisfied.

4 The Structure of Perceptual Experience: A New Look at Adverbialism

Imagine that you are looking at a red tomato. Your visual experience presents you with an object of determinate size, shape, and color. Of course, there may be nothing there—you may be hallucinating a tomato—or you may be looking at a green tomato in bad light. It is natural, at least for philosophers, to say that your visual experience *represents* a red tomato in front of you, and that the experience represents accurately only if there is such an object there; otherwise, it *misrepresents*. This is what *representationalism*,[1] the leading philosophical account of perceptual experience, would say. I don't think there is anything wrong with speaking this way; indeed, I think it is basically correct, but I don't think that representationalism provides the most illuminating account of perceptual experience. In this chapter, I defend an alternative proposal, inspired by traditional adverbialism about perceptual experience. The view, which I call *external sortalism*, allows us to retain what is unobjectionable about representationalism—that perceptual experiences have accuracy conditions—without its problematic commitments.

4.1 Orthodox Representationalism about Perceptual Experience

Representationalism holds not only that experiences have content but also that the content they have fully determines their *phenomenal character*—that is, what it is like for the subject to undergo the experience. So, for example, what it is like for you as you visually experience a ripe, red tomato is fully determined by the experience's content—that is, what the experience "says" about the way things are in front of you now. That content will only be imperfectly captured by a description used to characterize the experience. The experience has its content whether or not things are in fact that way, in other words, whether the experience is veridical or illusory. On this view, experiences are said to have their contents *essentially*—if a particular experience had a different content, it would be a different type of experience.

Representationalism comes in various strengths. *Intermodal* theorists hold that phenomenal differences among sensory modalities—the difference, say, between seeing a plane flying overhead and hearing it overhead—are determined by a difference in content. *Intramodal* representationalists deny this, holding that phenomenal differences among experiences of different modalities are to be explained functionally rather than solely in terms of content. *Unrestricted* representationalism extends the thesis to bodily sensations such as pains and orgasm, and *utterly unrestricted* representationalism holds that *any* mental state with phenomenal character, including emotions and moods, has content that fully determines its phenomenal character.[2]

Representationalists typically hold unrestricted versions of the thesis.[3] Byrne (2001) argues that restricted versions of representationalism are hard to defend:

> If there were an argument for [representationalism] about visual experiences, why couldn't it be adapted to the case of itches? Conversely, if a convincing counterexample to supervenience [of phenomenal character on content] were produced that would raise the suspicion that counterexamples concerning other experiences with propositional contents (visual experiences, as it may be) are waiting in the wings. (2001, 205–206)

Lycan (2015) appeals to a similar idea, pointing out that restricting the thesis to perceptual experience would be ad hoc. If the representational content of some mental states really determines their phenomenal character, then why should content not determine the phenomenal character of any mental state that has phenomenology? It is likely to be something in the general nature of both content and phenomenal character in virtue of which the former determines the latter. A restricted version of the thesis leaves the phenomenal character of the excluded states unexplained.

I find Byrne's and Lycan's arguments against restricted versions of representationalism convincing, but they suggest a strategy for undermining even the weakest version of representationalism[4]: focus first on types of experiences that are not *patently* intentional—say, bodily sensations and moods—and argue that their phenomenal character is *not* determined by content. Then, on the assumption that Byrne and Lycan are right that restricting representationalism to perceptual experience is indeed ad hoc and poorly motivated, use the argument against representationalism for bodily sensations and moods as "the thin end of the wedge," undermining the idea that content determines phenomenal character in the best case for representationalism—namely, perceptual experience. In the rest of this section, I offer such an argument. But my main interest in this chapter is to sketch what I think is a better alternative to representationalism as an account of perceptual experience.

So, I will focus first on bodily sensations and moods. Representationalists have proposed that the pain caused by spraining my ankle means, variously, *there is tissue damage in the ankle*, or the less specific *there is a disorder in the ankle* (Dretske 2003; Tye 1995, Lycan 1996), or *don't put weight on the foot* (Klein's [2015] "imperativist" proposal). These are distinct contents with different satisfaction conditions. One might wonder how this dispute is to be settled. What non-contentful facts ground the correct content attribution? Perhaps a future well-confirmed neuroscientific account of conscious mental states would support one of these content attributions over the others.[5] But the issue here concerns the *introspectable* content of these states, presumably something we can access simply by being in these states, and so it is not clear how the attribution of content to sub-personal states in cognitive science bears on the content of conscious experience.

There is something to be said for each of the competing content proposals. Pain attributions do not contain intentional verbs and are not completed by complement clauses (e.g., *that clauses*); nonetheless, pains certainly seem to be telling us something—*something is wrong with the relevant body part, there is tissue damage there, be careful not to stress it further*, and so on. But just because a pain (in the typical case) provides the subject with information about the state of her body and, moreover, tends to elicit protective behavior, it doesn't follow that the state literally has satisfaction conditions. The idea I propose is that characterizing pains and other bodily sensations in representational terms is simply a useful and natural way of talking about various functional properties of these states. For example, to say that a burning sensation in the finger represents tissue damage there is to emphasize the cause of the pain; to say that a sharp pain in the ankle has the imperative content *don't put weight on the foot* is to highlight the effects of the proscribed behavior (that putting weight on the foot will cause further pain and bodily injury).

In other words, to the extent that attributing content to pains and other bodily sensations is appropriate, it is because the attribution plays the role of a *gloss* on the complex functional role (or a salient aspect of that role) of these states. Similarly, for moods: perhaps euphoria can be said to represent general *wonderfulness* (Mendelovici 2014) and anxiety to represent *a departure from functional equilibrium* (Tye 1995). One may find these content attributions to moods to be a stretch, but, again, to the extent that they are appropriate, it is because they serve a variety of useful purposes—as with pains, to specify typical causes but also to suggest associated attitudes and emotions, perhaps to predict how the subject will respond to various stimuli, and so on. In general, I propose that we think of content attributions to pains and moods as playing a heuristic role, rather than construing them as attributing determinate satisfaction conditions to these states. Accordingly, the various representationalist proposals would be better construed as attempts to justify a particular type of contentful gloss.

Once the claim that content determines the phenomenal character of pains and moods is undermined, the general anti-representationalist argument then proceeds as follows: If Byrne and Lycan are right that restricting representationalism to perceptual experience is ad hoc and unmotivated, then a unified non-representationalist account that covers perceptual experience suggests itself. Rather than determining phenomenal character, content plays a merely heuristic, nonessential, role for *all* experience, including perceptual experience. In general, content attributions serve as a gloss on the complex functional role of experiences.

I don't want to put too much stock in this argument. Rather, let's return to our focus on perceptual experience. Despite representationalism's intuitive plausibility, there is a better alternative, which I sketch in the rest of the chapter. Eventually I will present a more direct argument for the idea that the attribution of content to perceptual experience, like the attribution of content to pains and moods, is best construed as a kind of *gloss*.

4.2 *External Sortalism*: An Adverbialist-Inspired Account of Perceptual Experience

So-called *adverbial* theories of perception, proposed originally by Ducasse (1942) as an antidote to sense data theories, rejected the idea that perceptual experiences are to be analyzed as relations between an experiencing subject and an entity of some sort. Sense data theories claim that a perceptual experience consists in a primitive relation—*sensing*—between a subject and a special sort of mental object—a *sense datum*. Naive realism holds that a veridical perceptual experience is essentially a relation between the subject and the perceived physical object. Typically, the view claims that the experience is *constituted*, in part, by the perceived object. (Naïve realists therefore must give a different analysis for hallucinatory experience. The resulting view goes by the name *disjunctivism*.) Representationalism holds that a perceptual experience is essentially a relation between the subject and what the experience is a representation *of*, say, the red tomato in front of me. Adverbial theories reject relationalist, or act/object, analyses altogether. According to adverbialism, perceptual experiences should be understood as *episodes of sensing in particular ways*.

The idea has traditionally been expressed as follows: *A subject S has a perceptual experience as of a property F* if and only if S senses *F-ly*. For example, S has a perceptual experience as of something red if and only if S senses *redly*. I follow Sellars (1956; 1975) and Tye (1984) in understanding *sensing F-ly*, canonically, as *S senses in an F manner*. For example, S has a perceptual experience as of a red tomato if and only if *S senses in a red tomato manner*. One can sense in an F manner whether or not the property F is in fact instantiated nearby, in other words, whether the experience is veridical or illusory. The name "adverbial theory" designates a *family* of views, because, as we shall see, "sensing in an F manner" can itself be understood in different ways.

An initial clarificatory remark is in order. "Adverbialism" is a rather unfortunate name for this family of views. Adverbial theories do not treat perceptual experiences as, literally, adverbial. *Adverb* is a linguistic category; it only applies to ways of describing experience. We can use adverbs to describe the ways that subjects experience the world, but the ways themselves are properties or states of experiencing subjects, not descriptions. The adverbial theory of perceptual experience is not a linguistic thesis, though early advocates were not always careful in their formulation of the view.[6] In any event, the view developed here, which is inspired by a particular type of adverbialism, is explicitly *not* a linguistic thesis, though it does, of course, have implications for how perceptual experience attributions are to be understood. I call the view *external sortalism* because it stresses the role of the external world in fixing the "ways" or "manners" that individuate experiential types. The view shares with traditional adverbialism a rejection of "act/object" analyses—the idea that experiences are relations of a special sort between experiencing subjects and an entity of some kind, whether a perceived external object, a content, a private mental object, or something else.[7]

One way that adverbialism has traditionally been developed is the "sortal" or "topic neutral" view (see Sellars 1956; 1975; Cornman 1971; and Tye 1975; 1984). According to this view, *S senses in a red square manner* is to be understood as follows: S is undergoing the type of experience that normally occurs when one is facing a red square in good light with eyes open, and so on. On this view, predicates such as "sensing in a red square manner" and "sensing in a green circle manner" provide a way of sorting experiences into equivalence classes based on their normal external causes—red squares and green circles, respectively. Once they are sorted, we can go on to think and talk about our experiences.

The "sortal" idea is developed by Wilfrid Sellars in his classic (1956) paper (sec. 61, 3). Here is Tye (1984) channeling Sellars: "Manners of sensing are conceived of as resembling and differing from one another in ways systematically analogous to the ways in which their 'corresponding' physical objects, i.e. their normal causes, resemble and differ from one another" (43).

Here is Sellars himself:

> The essential feature of the analogy is that: visual impressions stand to one another in a system of ways of resembling and differing that is structurally similar to the ways in which the colors and shapes of visible objects resemble and differ from one another. (1956, 61 [3])

The basic idea is that we use the external world as a *model* for sorting perceptual experiences into types, so that we can say, for example, that what I am having right now is a black cat type of experience. We can think of the view, schematically, as follows (figure 4.1).

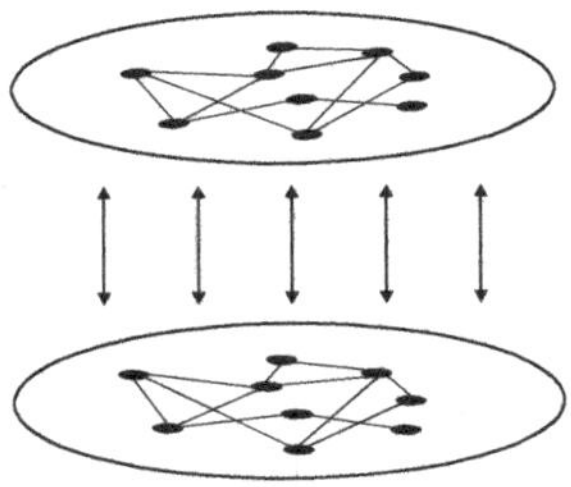

Figure 4.1
The external world as a *model* of perceptual experience.

Figure 4.1 depicts (at the top) the external world that serves as the model for perceptual experience and (at the bottom) perceptual experiences that the external world framework models. We can think of the matter, formally, as a mapping between relational structures, perhaps a homomorphism. It is in virtue of this shared structure that the external world framework (hereafter, S1 or EW) can serve as a model for conceptualizing, talking about, and reasoning surrogatively about perceptual experiences (hereafter, S2 or PE).

The terminology and general idea I appeal to here are from Swoyer (1991). Matthews (2007) develops and extends the idea, applying it to propositional attitudes. In surrogative reasoning, we use a model to reason indirectly but faithfully about relations in the modeled domain. The most common example is, of course, from physical measurement. We can reason about extensive physical magnitudes such as mass or temperature by reasoning about operations on the real numbers used to index or refer to particular quantities of the magnitude, as in "a mass of 5 kg" and "a temperature of 28°C."

It should be noted that when we talk about the "external world" in this modeling context, we must mean the external world as we perceive it—we have no other access to the external world except via our sensory-motor interactions with it. But what is relevant is the external world as we perceive it over the course of time. The basis for sorting experiences is the external world, abstracting away from particular experiences, regularized and stabilized by repeated perceptual and motor interactions with the environment, under different lighting conditions, from different perspectives, with difference interests and concerns in mind, and so on.

Much more could be said about the structure of the external world model, but the project here is not to specify its structure; rather, the point is that, whatever that structure is, the structure gets exploited in our thinking and talking about perceptual experience. There are likely to be aspects of the external world structure that do not get exploited in the application to perceptual experience, in the same way that reasoning surrogatively in the real numbers about physical magnitudes does not exploit all possible relations defined on the reals. (For example, the notion of one object being twice another is defined for length but not for temperature.) Representationalists about perceptual experience face questions about how rich the content of experience is; for example, are high-level causal relations represented in visual experience or are only more basic properties such as color, shape, and size represented? On the external sortalism account, such questions can be understood as questions about which aspects of the external world framework are exploited in the application to perceptual experience.

There are, of course, lots of questions about how we come, as individuals, to acquire our conception of the external world. The distinction between appearance and reality is something that emerges in early childhood. Developmental psychologists tell us that young children come to grasp the distinction between appearance and reality only at around age five, consistently failing appearance-reality (AR) tests at age three but consistently passing them at age six. In a well-known study, Flavell et al. (1986) show that three-year-olds do not have a conception of objects with stable color properties; at age six, they do. Children are shown a red car and asked what color it is. They respond "red." A filter is then lowered over the car, making it look black. The children are again asked what color the car is. Three-year-olds say "black"; by six they consistently say "red," indicating that at this age they know that objects have stable colors. Before grasping the distinction, and more importantly, the idea of a relatively stable external reality, children have no stable way of categorizing their experiential states. Flavell (1986) reports:

> It is noted that, while children of 6–7 years manage simple appearance–reality tasks with ease, they have great difficulty reflecting on and talking about such appearance–reality notions as "looks like," "really and truly," and especially, "looks different from the way it really and truly is." Children of 11–12 years, and to an even greater degree college students [!], however, give evidence of possessing a substantial body of rich, readily available, and explicit knowledge in this area. (418)

This strongly suggests that while six- to seven-year-olds have acquired the stable physical object framework, at this age, they have not yet fully acquired an understanding of perceptual experience (the "looks" framework). The latter is acquired only after the former, as required by external sortalism.

To think of the perceived external world in the terms suggested by external sortalism is to adopt an allocentric conception of the world, one in which among the various objects in that world will be experiencing subjects, including oneself, standing in relations to other objects. But this framework will also include ego-involving relations of the sort "my location being such that there is a chair in front of a table" that can be used to sort my perceptual experience as one of the type "there is a chair in front of a table." We can think of the external world framework as made up of possible *situations* or *states of affairs* constituted by objects and their perceptible properties and relations to each other.

Let me elaborate the key idea. Perceptible properties bear lots of relations to each other, for example, *set-theoretic (*determinable/determinate*) relations*—a scarlet car is a type of red car; *similarity relations*—a man's black shirt is more similar, along the color dimension, to a toddler's navy blue shirt than to a man's pink shirt, though more similar to the latter along the size dimension—and so on. Analogously, visual experiences of cars and visual experiences of shirts stand in roughly the same set-theoretic and similarity relations, respectively: a scarlet car experience is a type of red car experience, a man's black shirt experience is more similar along one dimension to a child's navy-blue shirt experience than to a man's pink shirt experience and more different from it along another. Our grasp of the similarity and difference relations that hold among experiences is parasitic on, and derivative from, our grasp of the similarity and difference relations that hold among objects in the world, as perceived by us.

We can be noncommittal about the entities modeled by the EW framework—perceptual experiences (S2 in figure 4.1)—beyond noting that they are perceptual experience *types*. They may be conceived of as properties or occurrent states of subjects, or, alternatively, as events, processes, or activities of subjects, though of course there may be independent reasons for preferring one or other of these views. And while the similarities and differences that hold among perceptual experiences are similarities and differences in phenomenal character, we will be noncommittal about the properties (intrinsic or relational) of experiences that account for phenomenal character. Ultimately, this is something for perceptual science to tell us about.

Following early proponents of "topic-neutral" adverbialism, external sortalism analyses *S senses in a red square manner* as follows: "S is undergoing the type of experience that normally occurs when one is facing a red square in good light with eyes open, and so on."

But the thesis requires an important qualification. As Jacobson and Putnam (2014) point out, it cannot be assumed that all subjects—even all *normal* subjects—looking at a red square in good light will undergo experiences with exactly the same type of phenomenal character. We know, for example, that judgments about unique colors differ across subjects. So, the expression "the type of experience" in the canonical formulation of external sortalism should be understood as designating a *range* of experiences with similar phenomenal characters. The view does not require that my experience of a given red square has exactly the same phenomenal character as yours; it requires only that our experiences be phenomenally similar.[8] It also requires that our experiences occupy roughly the same location in the relational structure. One may wonder how we know that our experiences occupy the same location in the relational structure. Overall similarity in our physical constitution makes it highly likely that they do, and the smooth communication facilitated by our shared language is evidence that they do.

Let me say more about how the external world framework is recruited for use in thinking and talking about perceptual experience. The sortals made available by the external world are applied to experiences by making use of descriptions that advert to what I call their *external world analogues*. We say (or think to ourselves) "I see a red tomato" or "I hear thunder." Normally, the point of such talk is to say something about the world, about tomatoes and thunder, but occasionally, especially if there is reason to think that things may not be as they seem, the focus is one's own experience.[9] An obvious but significant point is that the same form of words can do double duty, characterizing not only an external state of affairs but also a subject's mental state. In its latter usage, the description functions to pick out the experience by applying a predicate that specifies a particular location in a relational experience space (S2 in figure 4.1), a space that shares the same abstract structure as the relational space of the external world model (S1 in figure 4.1). We understand and can apply the scheme to experiences only because we are already familiar with external objects and their perceptible properties. The central idea of external sortalism is supported by the oft-noted *transparency* of perceptual experience, the idea that, typically, in perceptual experience, we are aware of the external world and not of properties of the experience itself.[10]

I have focused on visual experience, our richest and most informative sense. The extent to which external sortalism can be extended to the other sense modalities is an open question. Auditory experiences are similar to and different from each other along dimensions corresponding to the pitch, volume, and timbre of the sounds that typically cause them.[11] Haptic experiences are typically sorted along dimensions structurally similar to the texture, shape, size, and temperature of felt objects and substances.[12] Gustatory and olfactory experiences, by contrast, are typically sorted in terms of objects that typically cause them—*it smells like a skunk, it tastes like cilantro*. We can articulate structure among tastes along at least four dimensions—sweet, sour, bitter, salty—abstracting away from the substances that instantiate these properties, but this ability likely develops quite late in children, after the experiences have been anchored to the external objects that cause them. In any event, there is currently little evidence of a relational space of external objects and properties (EW) structurally similar to taste and smell experiences (PE).[13]

Returning to visual experience, the description made available by the experience's external world analogue tells us quite a bit about what the experience is like. Once again, the key idea is that a visual experience as of a red tomato (that is, an experience of the sort that normally occurs when one is facing a red tomato in good light with eyes open) is like a visual experience as of a red fire engine (an experience of the sort that normally occurs when facing a red fire engine) in a particular way and unlike it in others. Our facility with the external world framework allows us to use it to reason surrogatively about our inner mental states, enabling us to predict and imagine what it would be like to have other experiences. Even though I have never seen a pink elephant, I can predict (and imagine) what the experience of seeing one would be like. This doesn't work as well for smells and tastes, but then our ability to think imaginatively about smells and tastes is more limited.

The sortings made available by the external world model are probably not univocal. Most times, a round coin on a table seen from an oblique angle looks like a round coin on a table; at other times, perhaps when we are looking at the coin in reduced light, without knowing it is a coin, it may appear ovoid. It may appear ovoid in circumstances where situational features—the subject's location with respect to the perceived object—are particularly salient. And we may be able to think ourselves into a situation where the coin appears ovoid by thinking about the 2D projection of the coin on a plane at an oblique angle. This ability is cognitively quite sophisticated and likely develops only quite late.[14]

External sortalism takes no stand on whether the similarities and differences among perceptual experiences, in particular, the similarities and differences among their phenomenal characters, hold in virtue of some intrinsic property of experiences. It is simply silent about this. In particular, the view does not claim that the phenomenal character of a red square experience is constituted by the conjunction of a tokening of a phenomenal property that we might call *e-red* (for *experiential red,* to distinguish it from *red,* a property of external objects and surfaces) and a tokening of a phenomenal property *e-square* (for *experiential square,* to distinguish it from *square,* a property of external objects).[15] There is no commitment to phenomenal properties or *qualia*. There is certainly no presumption that a visual experience of a red square is itself similar to the external object to which it is mapped, what I am calling its *external world analogue,* in this case a red square. Indeed, it is hard to even make sense of this idea. Berkeley made the point forcefully in his *Essay toward a New Theory of Vision* (1709/2009). So, the view does not purport to explain the phenomenal character of experience—that is, its distinctive phenomenology. The sorting provided by the external world framework gives us similarities and differences among phenomenal characters; it says nothing about the nature of phenomenal character itself.

We might wonder what external sortalism purports to explain, if not the phenomenal character of perceptual experience. External sortalism gives an account of our thought and talk about perceptual experience, as it functions in our ordinary lives. In virtue of sharing overall structure, the external world framework is recruited as a model for conceptualizing, talking, and reasoning about the internal world of experience. External sortalism explains how we acquire and develop the capacity to think and talk about experience, but beyond a commitment about its structure, it is silent about its intrinsic nature.

That external sortalism lacks the resources to explain the phenomenal character of experience may seem to be a serious failing of the view. Howard Robinson (personal communication) has raised this point as an objection to external sortalism. Fish (2010, 45) seems to concur, claiming that the so-called qualia version of adverbialism is better placed than sortal versions to explain phenomenal character.[16]

Let us pause for a moment to compare these two versions of adverbialism. Qualia adverbialism is sometimes taken to be an explicit articulation of the original adverbialism thesis, the most natural way to understand the locution "sensing redly," which was introduced originally to make the point that perceptual experiences are not relations between experiencing subjects and objects of some sort. Here are Crane and French (2021):

> Adverbialism theory is committed to the view that experiencing something white, for example, involves one's experience being modified in a certain way: experiencing whitely. A natural way to understand this is that the experience is an event, and the modification of it is a property of that event. Since this property is both intrinsic (as opposed to relational or representational) and phenomenal then this way of understanding the adverbial theory is committed to the existence of qualia. (2021, sec.3.2)

Qualia adverbialism is not a natural construal of the sortal or "topic-neutral" thesis advocated by Sellars (1956; 1975),

Cornman (1971), and later Tye (1975; 1984). Such theorists explicitly eschewed appeal to mental objects (for example, sense data) and mental properties such as qualia. Nonetheless, qualia theory is the most popular contemporary version of adverbialism (see, for example, Kriegel 2008). Crane (2000) agrees with Robinson and Fish that a qualia analysis of "sensing redly" gives the adverbialist the best chance of explaining the phenomenal character of experience:

> The most natural interpretation would take this property to also be a phenomenal or qualitative property that explains why it is like this for Jones. On this natural reading of adverbialism, then, the adverbialist . . . "holds the qualities sensed in experience to be modifications of experience itself." (177)

Returning to the objection that external sortalism lacks the resources to explain the phenomenal character of experience: if this is a failing, it is not specific to external sortalism. No *philosophical* theory of perceptual experience—not representationalism, sense data theory, naïve realism, nor the qualia version of adverbialism—explains phenomenal character. None of these theories explain why looking at a red tomato produces an experience with its distinctive phenomenal character rather than, say, one with the phenomenal character characteristic of a visual experience of green grass, or a mental state with no phenomenal character at all. Rather, they, like external sortalism, simply take these facts for granted. In other words, none of these theories close the famous "explanatory gap" (Levine 1983).

To sharpen the point, it is useful to focus on sense-data theory for a moment. Proponents (e.g., Broad 1923; Robinson 1994) claim to have an explanation of the distinctive phenomenal character of experience in virtue of the theory satisfying what Robinson (1994) calls the *Phenomenal Principle,* which holds: "If there sensibly appears to a subject to be something which possesses a particular sensible quality then there is something of which the subject is aware which does possess that quality" (Robinson 1994, 32).

The idea is that a visual experience as of a red tomato has its distinctive phenomenal character in virtue of the subject bearing a sensing relation to something—a sense datum—that is *itself* red and tomato-shaped. Moreover, the correspondence is claimed to explain the experience's distinctive phenomenal character.

There are several problems with the proposal. In the first place, external physical objects and private mental objects are unlikely to have *the very same properties* (that is, colors, shapes, sizes, and so on). Secondly, and more generally, it is not at all clear how positing an inner object with just the properties sensed in the experience provides an *explanation* of the experience's phenomenology. To posit such an inner object as an explanation is akin to appealing to a *dormitive virtue* to "explain" the sleep-inducing properties of a drug. A similar point applies to the explanatory potential of invoking *qualia*—which just *are* the phenomenally salient properties of experiences—to explain phenomenal character. Looking at a red tomato typically produces in a subject a perceptual experience with a particular quale—*e-red*—but appealing to the quale is merely a way of specifying the experience's phenomenal character and distinguishing it from experiences with *e-green* qualia. It is not an explanation of *how* or *why* the experience of looking at a tomato has that particular phenomenal character. Qualia theories, in respect of their potential to explain phenomenal character, are on a par with external sortalism. The appeal to qualia is a wheel turning nothing; it simply attaches a label to what we would like explained.

I said at the beginning of the chapter that there is nothing wrong with saying that a red tomato experience *represents* a red tomato. The mistake is in thinking that this characterization *explains* the phenomenal character of the experience. Representationalism—like its rivals naïve realism, sense data theory, and qualia adverbialism—says nothing substantive about why a red tomato experience has the particular phenomenal character that it has, rather than some other phenomenal character, or none at all. In this respect, these theories are on a par with external sortalism, which has no such explanatory pretensions.

Phenomenal character, if it is to be explained at all, will be explained by the perceptual and cognitive sciences. As I have said, external sortalism aims to do something quite different—namely, explaining our thought and talk about perceptual experience as these function in our ordinary lives. A third project, served by a "philosophical theory of perceptual experience," often involves philosophers getting out their tool kits—some philosophers prefer representations, others like sense data—and characterizing perceptual experience in their favored terms. This can be illuminating, as long as we don't think that phenomenal character has thereby been explained.

There are several advantages to thinking about perceptual experiences in the way suggested by external sortalism. In the first place, the view involves no specific metaphysical commitments about the mental—it posits no distinctively mental objects and, as I have emphasized, says nothing about the intrinsic properties of experience—and is therefore compatible with either a materialist or a dualist metaphysics. Proponents of early sortal versions of adverbialism made a similar point about their view and claimed it as an advantage over sense data theory, with the latter's commitment to metaphysically problematic private objects.[17] As noted earlier, external sortalism is also noncommittal about whether experiences are best construed as properties, states, or activities of subjects.

Secondly, and more importantly, the view gives us a plausible explanation of how children learn to think and talk about experiences, which are, after all, episodes *private* to the experiencing subject, where by this I mean not that the subject has authoritative or privileged access to them, as these notions are generally understood, but simply that my experiences are private in the sense that they are *mine*. External objects and their properties are publicly accessible, and although the details are complex, there is no inherent mystery about how we learn to talk to each other about them. Our ways of thinking and talking about our experiences are parasitic on public object thought and talk. As Sellars (1956) pointed out, reports of mental states have a dual character: (1) they are reports of episodes that are private, in the above sense, and (2) they are part of public discourse and hence intelligible to others. External sortalism makes this duality less puzzling.[18]

External sortalism differs from adverbialism, as traditionally understood, in several important respects.

1. As noted previously, it is not a linguistic thesis about the logical form of perceptual experience attributions. Hence the name *adverbialism* is misleading. However, in analyzing perceptual experience attributions as follows: "S is having a perceptual experience as of a red square if and only if S is experiencing in a red square manner" and this, canonically, as "S is undergoing the type of experience that normally occurs when one is facing a red square in good light, with eyes open, and so on," the terms ("red," "square")

that specify the "manners" have their ordinary meaning. Compositionality of experiential attributions is therefore preserved. The "manners" so specified can be arbitrarily complex.

2. Moreover, external sortalism, unlike some versions of adverbialism, as we shall see in the next section, has no trouble accounting for the validity of such inferences as "S is having a red square experience, therefore S is having a red experience." Experiences that normally occur when one is facing a red square (with eyes open, and so on) are a proper subset of experiences that normally occur when one is facing something red (with eyes open, and so on). We get these inferences from set-theoretic relations. And the set-theoretic relations that hold among perceptual experiences are grounded in the set-theoretic relations that hold among their external world analogues: the set of red, square things is a proper subset of the set of red things. So the determinable/determinate relations that hold among perceptual experiences are not brute.
3. External sortalism does not construe the modifier in perceptual experience attributions (the *manners*) as modifying (that is, as attributing some property to) experiences in the way that traditional adverbialism is often understood to do (e.g., redly, squarely). As I have stressed, what is attributed to the experience is a location in a relational structure, not any intrinsic properties that might explain why these similarities and differences hold. As we shall see in the next section, this point is significant for external sortalism's potential to avoid a notorious problem faced by traditional versions of adverbialism.

4.3 Objections and Replies

4.3.1 The Many Properties Problem

Objection: Adverbial theories of perceptual experience suffer from what Jackson (1975) called the *many properties problem.* Imagine that you are looking at a red square and a green circle. According to adverbialism, you are sensing *redly, squarely, greenly,* and *circularly.* But an experience of a red circle and a green square is also an instance of sensing *redly, squarely, greenly,* and *circularly,* yet of course it is a different type of experience. In rejecting the idea that perceptual experience is a relation between a subject and objects of perception, adverbialism appears to lack the resources to distinguish different ways that sensed properties can be bound together in distinct types of experiences. Hence, the problem is sometimes called *the binding problem.*

Reply: According to external sortalism, to sense in a red square and a green circle manner is to undergo the type of experience that normally occurs when one is facing a red square and a green circle, in good light with eyes open, and so on. To sense in a red circle and a green square manner is to undergo the type of experience that normally occurs when one is facing a *red circle* and a *green square.* The external situations or states of affairs that normally cause the two experiences and that provide the basis for sorting the experiences into types are different. These situations occupy distinct locations in the relational structure (S1—the EW framework) that serves as a model for conceptualizing perceptual experiences. The fact that both experiences can be described as a sensing *redly, squarely, greenly,* and *circularly* is irrelevant; this is not the canonical characterization of the experiences given by external sortalism. External sortalism does not face a binding problem because, unlike the qualia version of adverbialism, external sortalism does not treat perceptual experiences as constituted by conjunctions of phenomenal properties or qualia—in this case *e-red, e-square, e-green, e-circular*—that correspond to the perceptible properties of perceived objects and that must be bound together in the right way. The external states of affairs that provide the basis for sorting experiences are constituted by observable objects that come with their perceptible properties bound together "for free," as it were. The *world* binds them together in one way rather than another. The many properties problem simply does not arise for external sortalism. It has a substantial advantage over other versions of adverbialism.

Traditionally, adverbialists have considered some moves intended to mitigate the many properties problem, but these moves raise other problems for the view. One strategy is to "fuse" the predicates characterizing the distinct experiences, treating the experience of a red circle and a green square as a *sensing redly-squarely-greenly-circularly* and the distinct experience of a red circle and a green square as a *sensing redly-circularly-greenly-squarely*. This strategy distinguishes the two experiences, but it has several unfortunate consequences. The meaning of a fused predicate is not determined compositionally from the meanings of its constituents; it *has* no meaningful constituents. Compare with external sortalism: as I noted earlier, the terms that specify the manners in an arbitrarily complex experiential attribution have their ordinary meanings, and hence compositionality is preserved. Another unfortunate consequence of the fusion strategy is that the view cannot account for the validity for such inferences as "S is having a red square and a green circle experience, hence S is having a red square experience," an inference that on the external sortalism view is accounted for by appeal to set-theoretic relations that hold among experiences, relations that are grounded in set-theoretic relations that hold among their external world analogues.

Kriegel (2008) appeals to determinate/determinable relations among modes of sensing—among *qualia*—to address the "many properties" problem, but unless these relations are parasitic on external object and property relations, as they are for external sortalism, they are problematically brute. See Grzankowski (2017) for criticism of Kriegel (2008).[19]

4.3.2 The Many Relations Problem

Objection: A recently noted variant on the many properties problem (see Dinges 2015) is thought to be particularly challenging for adverbial accounts. Imagine that you are looking at a red object that is brighter than a green one and to the right of a blue one. The challenge for the adverbialist is to distinguish this experience from an experience of a red object that is brighter than a green one, but where it is the green object that is to the right of the blue one. It isn't clear that adverbialism has the resources to distinguish experiences where the relations are distributed differently among the objects.

Reply: External sortalism handles multiple relations in the same way as multiple properties. In the imagined case, I am undergoing the type of experience that normally occurs when I am facing a red object that is brighter than a green one and to the right of a blue one (with my eyes open, and so on). This is a distinct experience from the type of experience I have when facing a red object that is brighter than a green one where it is the green object that is to the right of the blue one. The two states of affairs are distinct; they occupy distinct locations in the external world framework (S1, EW); analogously, the two experiences occupy distinct locations in the experiential space (S2, PE) modeled by the external world framework. Once again, the external world does the work of grounding the differences between distinct types of perceptual experiences.

4.3.3 External Sortalism Is Problematically Circular

Objection: The external world framework (EW) is acquired from perceptual experience. External sortalism claims that we then go on to conceptualize perceptual experience in terms of the EW framework. But this looks to be circular—we are sorting perceptual experience in terms of itself.

Reply: Infants discriminate, via perception, red things from green things, squares from circles, and so on. Later, they go on to conceptualize the similarities and differences among physical objects in terms of stable properties such as color, size, and shape. This process involves abstraction and generalization from what is learned from particular experiences, but as the earlier discussion makes clear, the external world framework is acquired without the child making use of mentalistic notions such as "experience" or "the way things look." Finally, children go on to apply the EW framework to their mental states, characterizing them in ways structurally similar to the ways that they think of the objects and states of affairs that make up the EW model. There is no circularity here. As the above discussion of the appearance/reality distinction strongly suggests, the perceptual experience framework (S2 in figure 4.1) is not acquired until after the child has mastered the external world framework (S1) to which it is structurally similar. Thus, external sortalism provides at least a partial explanation of the transparency of perceptual experience, the fact that when we focus on what a perceptual experience is like we typically do not discover properties of the experience itself but rather properties of the perceived external world.

4.3.4 External Sortalism Doesn't Account for the Possibility of Inverted Spectra

Objection: In characterizing perceptual experience in terms of its normal external causes, external sortalism doesn't leave room for the possibility that two subjects may be spectrally inverted with respect to each other. External sortalism treats two subjects undergoing a perceptual experience caused by looking at something red as undergoing the same *type* of experience even though one subject may in fact be having an experience phenomenally indistinguishable from the experience that the other has when looking at something green. The view seems to give the wrong verdict about inverted spectra cases.

Reply: External sortalism analyzes *S senses redly,* or *S senses in a red manner* as follows: S is undergoing the type of experience that normally occurs when one is facing something red in good light with eyes open, and so on. If S is an inverted subject, then she has that type of experience when she (S) is facing something *green*. But that just *is* the type of experience that normal subjects undergo when facing something red. (Recall the earlier qualification that the expression "the type of experience" in the canonical formulation of external sortalism should be understood as designating a *range* of experiences with similar phenomenal characters.) So, external sortalism gets inverted spectra cases right.

External sortalism characterizes perceptual experience by tacking it down to the types of experiences had by subjects like us when in causal contact with publicly observable objects and properties. This apparatus allows us, as subjects of perceptual experience but also as philosophical theorists of perceptual experience, to refer to a class of phenomenally similar perceptual experiences, while saying nothing about what gives perceptual experiences their phenomenal character. This same apparatus allows the formulation of the inverted spectra scenario: inverted subjects differ from normal subjects only in very specific and systematic ways, and those ways are characterized by reference to publicly observable objects and properties (e.g., looking at *something green*). And so, external sortalism handles inverted cases smoothly.

External sortalism has nothing to say about subjects whose perceptual experiences are radically different from our own. There is no reason to think that different species or alien life forms will satisfy the left-hand side of the external sortalism schema, that they will share our *ways* or *manners* of sensing. There is no reason to think that they will sense in a red square manner—that is, that they will have the sort of experience that we have when looking at a red square. It is appropriate that the theory is silent about such subjects; we have no idea what it would be like to undergo their experiences. But whatever their perceptual experiences are like, it is highly likely that they can be sorted in terms of the external objects and properties that are observable for them, whatever these objects and properties are, thus enabling them to think, talk, and reason surrogatively about their experiences, as we can about ours. In other words, the general picture and underlying motivation for external sortalism, in particular, the idea that the external world can serve as a model for perceptual experience in virtue of the shared structure of the two domains, will apply to alien experience as well. This suggests a methodology for beginning to investigate alien experience: attempt to characterize the framework in virtue of which aliens conceptualize the objects and properties that are observable for them. On the plausible assumption that this framework shares a relational structure with their perceptual experiences and so provides a model for their thinking, talking about, and reasoning surrogatively about their experience, as our external world framework does for us, we can get a glimpse of the structure of their inner lives.

4.4 The Role of Representational Content in External Sortalism

So far, nothing has been said about experiences having *representational content*. In understanding perceptual experiences as episodes of sensing in particular ways, traditional adverbialism does not construe perceptual experiences as representational—that is, as having content. Fish (2010) characterizes the two general types of adverbial theories—sortal and qualia theories—as explicitly rejecting the idea that perceptual experiences are representational. Though Siegel (2016, sec. 2.1) observes that

> so long as the core thesis of adverbialism is that experiences are modifications or properties of the subject, adverbialism leaves open whether experiences are assessable for accuracy or not. For all that core thesis says, being appeared-to F-ly could be a way of representing that something is F.

In other words, adverbialism leaves open whether experiences have representational content.

To have content is to be assessable for accuracy. As Crane (1992) puts it,

> To say that a state has content is just to say that it represents the world as being a certain way. It thus has . . . a "correctness condition"—the condition under which it represents the world accurately. (1992, 139)

An adequate philosophical account of perceptual experience should account for the fact that we sometimes classify perceptual experiences as accurate or inaccurate. We talk about Macbeth hallucinating a dagger, a red car looking black in dim light, and so on. External sortalism can account for this feature of our ordinary practice, though it denies (1) the representationalist claim that content determines the phenomenal character of experience, and (2) the idea that having accuracy conditions (i.e., content) is a deep, metaphysical fact about experiences, in other words, that it plays a fundamental role in type-individuating experiences. So the view comports very naturally with the deflationary account of representation that I develop in earlier chapters.

Moreover, orthodox representationalism requires an account of *how* perceptual experiences get their content. External sortalism already has such an account. Recall the earlier point that sortals are applied to experiences by making use of descriptions that advert to their external world analogues, descriptions such as "I see a red tomato." We can take the description to specify a *content*—an accuracy condition—that is satisfied just in case there is a red tomato in front of one causing the experience. To make this move is to treat perceptual experiences as *representing* (and occasionally *misrepresenting*) the objects and properties in the external world model that provide the basis for classifying them. This is to take a description whose primary function is to characterize the experience in terms of its location in a relational space to, in addition, specify an accuracy condition. Suppose that I am currently undergoing the type of experience that normally occurs when facing a red tomato in good light with eyes open, and so on. I can consider whether on *this* occasion there actually is such an object in front of me causing the experience. If there is, then the experience accurately represents; if not, it is a misrepresentation.

Perception, of course, is our principal way of cognitively accessing the world and is the basis for many of our beliefs about the world. Construing perceptual experiences as having content outfits them to serve our epistemological purposes, a role played by content attribution in both scientific and commonsense psychology. We can consider whether my red tomato visual experience—the experience characterized by reference to its external world analogue (a red tomato with a particular size, shape, and so on)—justifies my current belief that there is a red tomato of a particular size and shape in front of me.

The way of thinking about content suggested by external sortalism is well motivated. It is highly likely that it is the overall relational structure of experience that makes a representational construal possible. There is no basis for positing a *resemblance* relation between experiences and their external world analogues, a relation that, were it to hold, might serve to underwrite a claim that a given experience, considered in isolation from other experiences to which it bears various similarity and difference relations, represents its external world analogue. It is the structural similarity between external objects and their perceptible properties on the one hand and perceptual experiences on the other that underpins our representational capacities—that makes us, *qua* perceivers, generally reliable about the world.[20] It is only against this background structure that we can assess individual locations in the relational space of experiences for accuracy. Without a *system* of shape and color experiences, no single shape or color experience has content or is accessible for accuracy; without a system of shape and color experiences, an experience would not count as a shape or color experience at all. The relational structure of experience grounds its representational status, as external sortalism makes explicit.

4.5 Summary and Conclusion

To recap: (1) The predicates made available by the external world framework enable us to conceptualize, talk, and reason surrogatively about our perceptual experience. They give us a *handle* on something that is private—the phenomenal character of experience—by situating the experience in a relational space of similarities and differences made available to us by the external world model (S1 in figure 4.1). My experience is private in the sense that it is *mine*, but not that it cannot be had by others. Many of you have had experiences phenomenally similar to that which I have when facing a red tomato.

(2) The predicates can be treated as *content attributions*, in other words, as specifying accuracy conditions, allowing me to evaluate whether, for example, on this occasion my sensing in a red tomato manner is actually caused by a red tomato—the experience's *external world analogue*—and so warrants the belief that there is a red tomato in front of me. This is an option that has an extrinsic motivation—namely, serving our various epistemological purposes; it is not required for understanding perceptual experience itself.

(3) On the view I am defending, content neither determines phenomenal character, as representationalists claim, nor explains it. The content attribution simply *characterizes* it, indirectly, by picking out the experience's location in a relational space structurally similar to that occupied by external

objects, as presented in perceptual experience. And this job can be done without construing perceptual experience as representational or contentful, simply by deploying external sortals—that is, by appeal to external world analogues.

I have identified a further respect in which external sortalism differs from traditional adverbialism: there is a sense in which experiences do bear a relation to their external world analogues—to their *intentional objects*, as representationalists might put it—but there is nothing "metaphysically weighty" or substantive about this relation. We can call it a *representation relation* if we like, but this is to be understood in a very thin sense.

In conclusion, attribution of content to perceptual experience, like the attribution of content to pains and moods, as suggested earlier, is best construed as a kind of *gloss*. Attributing content to perceptual experience provides a way of adverting to the experience's external world analogue that plays an essential role in characterizing the experience. Content plays a complex heuristic role in our various expository and epistemic practices but should not be construed as an essential or metaphysically deep property of experience.

Afterword

A recurring theme in these lectures is that characterizing mental states and processes as *representational* involves a kind of modeling. Computational psychology and computational neuroscience model the causal processes underlying mental capacities as mathematical processes; commonsense psychology models beliefs and the other attitudes as sentences, and the causal relations holding among the attitudes as inferential processes defined over sentences; perceptual experiences are modeled by the framework constructed out of the worldly objects and properties that typically cause them.

The linguistic model is particularly potent. We need to be wary of a natural tendency to read, without sufficient justification, properties of a model into the modeled domain, in this case taking mental phenomena to literally have the semantic properties—meaning, reference, and truth conditions—of the sentences used to attribute them. The deflationary program, with its emphasis on the pragmatic motivation for the attribution of mental content, is intended to be a corrective to that tendency.

The focus on pragmatic motivation may give rise to the suspicion that the deflationary account is really a version of *fictionalism* about mental representation. Let me allay that suspicion. Toon (2022) characterizes fictionalism as follows: "Mental fictionalism . . . claims that mental representations are useful fictions. People do not really have representations inside their heads, but talking as if they do helps us to make sense of their behavior" (52).

Mental representations, on the deflationary view, are *not* fictions; they are causally efficacious states or properties of subjects, inside their heads, as it were. Recall that characterizing something as a representation presupposes a distinction between a *vehicle* and the *content* that it carries, where the former has causal powers. A burden of these chapters has been to characterize *what it is* that is glossed by the attribution of representational content in each domain—states or structures posited in computational accounts of cognitive capacities, causal complexes appealed to in our ordinary practices of predicting, explaining, and regulating behavior, and finally, perceptual experiences. These are the vehicles that are assigned content in a pragmatically motivated intentional gloss. To be a mental representation on the deflationist account is to be a causally efficacious state or structure—in the case of commonsense psychology, a rather complex functional state—that is glossed as having semantic and intentional properties. To say that a representational characterization is a kind of modeling is not to imply that it is a fiction. The model is intended to be explanatory, to highlight real properties of the modeled system that play central roles in the explanation and prediction of behavior and other mental states. It is neither a fiction nor, as some fictionalists (e.g., Toon 2023) would have it, a metaphor.

Notes

Chapter 1

1. See Colombo and Series (2012) for a defense of this view. For a general discussion of the issue, see Zednik and Jakel (2016).

2. Chirimuuta (2021; 2024) also claims that computational characterization is metaphorical, but her view is much more subtle and interesting that either Bennett and Hacker's or Barrett's. She stresses the usefulness of the computational scheme for prediction and simplification but cautions "that we should not fall into the habit of thinking that the brain is literally a computer because the disanalogies between these two kinds of things cannot be discounted" (2024, 30–31). She cites abstraction and simplification as important features of computational modeling. But there will always be disanalogies between a model and the modeled domain. Abstraction and simplification, in particular, are pervasive features of modeling in science and are not sufficient to support a metaphorical interpretation of a computational characterization.

3. Craver (2006) draws a distinction between "how possibly" and "how actually" models. The former sketch how a capacity might possibly be accomplished, with no explanatory pretensions about how it is actually done.

4. Readers will recognize the similarity to Cummins's (1989) "tower-bridge" idea.

5. The examples I have chosen to illustrate the strategy are of relatively low-level phenomena, such as perceptual processing and motor control and insect cognition. This is no accident. The conspicuous successes of computational theorizing have been in characterizing highly modularized, informationally encapsulated processes such as those responsible for early vision and syntactic and phonological processing. The processes underlying high-level capacities such as decision-making and rational revision of belief have so far resisted computational characterization. Their intractability is due in part to the fact that general constraints on the information that may be relevant to decision-making and belief revision are exceedingly difficult to specify, making the attempt to specify the function computed by the system very challenging. As Fodor (2000) notes, just about anything may be relevant to such processes. He concludes that the prospects for a computational treatment are dim. It remains to be seen whether Fodor's pessimism is justified. The development of *large language models* (LLMs) illustrate the problem. Computational explanation aims at making cognition *intelligible*; success requires model transparency, which involves specifying the functions the system computes, the algorithms used, and so on. Given the huge datasets on which LLMs operate and their massive connectivity, they lack the sort of transparency that facilitates understanding and so are often said to be *black boxes*.

6. To preserve the way we normally talk, a miscomputation may be seen as a special kind of malfunction. It is also, as all mistakes are, a partial exercise of a competence (Ryle 1949).

7. Discussion of place cells is ubiquitous in the neuroscience literature. For influential early work, see O'Keefe (1976), O'Keefe and Dostrovsky (1971), O'Keefe and Nadel (1978), and Ranek (1973).

8. Intentionality is the power of mental states to represent things outside themselves. See Jacob (2023) for a useful general discussion.

9. See Dretske (1981) and Fodor (1990) for the most developed information-theoretic accounts.

10. See Millikan (1984), Papineau (1987; 1993), Matthen (1988), Neander (1995; 2017), Ryder (2004), Shea (2018), and many others for various teleosemantic proposals.

11. This is Neander's (1995) proposal. Her (2017) proposal—*moving wormlike stimulus*—is essentially the same idea.

12. Shea (2018) proposes *nutritious flying object* as the structure's content. This idea includes aspects of proposals (b) and (c).

13. See Egan (2020) and Egan (2022) for further discussion of Shea (2018) and Neander (2017), respectively.

14. Eric Schwitzgebel has made this point in conversation.

15. Following convention, I use the upper case to denote uninterpreted structures.

16. An anonymous referee has suggested that my account of the theory proper is similar to Cummins's (1989) account of physical computation as *program execution*, which Cummins explicates as *disciplined* (that is, *rule-governed*) *step-satisfaction*. But the two are quite different. A computational model of a capacity may characterize a computational process in terms of disciplined step-satisfaction. If it does, then that characterization will be an element of the theory proper, but many computational models—for example, Seung's (1996; 1998) model of oculomotor memory and Ullman's (1979) structure-from-motion mechanism, both described earlier—do not. See Shagrir (2022) for a critical discussion of Cummins's account of computation as step-satisfaction.

17. Cognitive content, though determined in part pragmatically, is still content, and the structures that carry it must be capable of misrepresenting, if only in the intentional gloss. Once structures are assigned content—even if that content assignment depends on the interests of theorists—then it is an objective matter on a given occasion whether the structure is tokened in the presence of the relevant condition. Suppose, for example, that given theorists' goal of explaining how the frog's visual system works, they assign to structure S the job of representing a small, dark moving thing (Neander's 1995 content proposal). Then, if S fires without the presence of a small, dark moving thing in the environment, it will misrepresent.

18. See the references in note 7.

19. See Facchin (2023) for detailed argument that, though neural maps may appear, to our eyes as theorists, to be structurally similar to

external features that they are said to represent, the required similarity may not really be present, and moreover, that even if the similarity were present, it would not play the required representational role. If Facchin is right, then representationalist accounts of neural maps commit a version of what Slezak (2023) calls *the spectator error*, interposing the theorist—that is, ourselves—between mind and world to adjudicate the reference of mental states.

20. For a skeptical take on the "cognitive map" idea, see Luo, Mok, and Love (2024). They challenge the assumption that cells with readily interpretable receptive fields (place cells, head-direction cells, border cells, and so on) underlie spatial cognition, arguing that what are commonly referred to as "spatial cells" may be "inevitable" products of general computational mechanisms and are not intrinsically spatial. Their studies suggest that such cells could manifest in any sufficiently rich information processing system. "Not only do we find spatial signals within a non-spatial system but also . . . these purported signals, ostensibly representing spatial knowledge, do not appear to have a privileged role for tangible downstream tasks related to spatial cognition as previously claimed" (p. 4 preprint).

21. Karen Neander (in her comments on my *Society for Philosophy and Psychology* keynote address at Duke University in June 2015), Ramsey (2020), Hutto and Myin (2020), and Newen and Vosgerau (2020) make versions of this objection.

22. There is, of course, a question about the metaphysical status of abstract objects, but this issue is independent of the dispute about how to understand talk of representation in cognitive science.

Chapter 2

1. Robust representational realists include Fodor (1987; 1990), Dretske (1981; 1995), Millikan (1984; 1989), Neander (2017), Papineau (1987; 1993), Shea (2018), and many others.

2. I will explain the notion in due course.

3. So, the requirement appears to rule out the view defended in Burge (2010).

4. See Sprevak (2013) and Toon (2022; 2023).

5. Hutto and Myin's (2012; 2017) *radical enactivism* program is probably best construed as a type of eliminativism. Motivated by what they call *the hard problem of content*, they eschew appeal to mental representation, arguing that the naturalistic resources of information invariance and teleological function are insufficient to account for content. Representation comes into play only when more basic cognition is supplemented by sociocultural, primarily linguistic, capacities.

6. Cash (2008) develops an interesting proposal that I will not discuss. He appeals to evolution to explain how the normativity that is essential to semantics may have arisen in a prelinguistic world. In criticizing reductive teleosemantic approaches, he argues that "content is not an intrinsic feature of the world, but a feature of our explanations of the world" (115), a thesis that is also central to my deflationary account of mental representation.

7. See the references in note 1. See Jacob (2023, sec. 9) for an overview of the project and some of its problems.

8. See the discussion in chapter 1 of Neander (2017) and Shea (2018), and see Egan (2020; 2022) for criticism of these two accounts.

9. See Horgan and Graham (2012), Kriegel (2013), Mendelovici (2018), and Mendelovici and Bourget (2014), among others.

10. I am assuming that public language *has* determinate reference; that, of course, is an independent issue. If it does not, then indeterminacy in public language will infect content attributions in cognitive science. This possibility is not of particular concern on the deflationary account since such attributions serve only various heuristic functions. If indeterminate linguistic content can serve our sundry purposes in ordinary life, there is no special reason to doubt that it can serve heuristic functions in science.

11. Under idealization, of course, since no finite system can compute a function defined on an infinite domain.

12. My function-theoretic notion is not to be understood as *functional* or *functionalism* in any sense of those notions.

13. As Papayannopoulos et al. (2022) note, computationally characterized systems are subject to *two* kinds of indeterminacy. The indeterminacy discussed by Facchin—what Papayannopoulos et al. call *intentional indeterminacy*—involves holding constant the system's computational vehicles—that is, the s_i characterized by the bottom mapping in figure 1.2 (the *realization function*)—and then varying the content (that is, the function that the system computes) assigned by the *interpretation function* f_I. A distinct type of indeterminacy—*functional indeterminacy*—noted by Shagrir (2001; 2022), involves carving up physical states of the system differently to yield a different taxonomy of computational vehicles s_i. Functional indeterminacy depends on the existence of multiple realization functions f_R. (Thanks to an anonymous referee for pressing me to note this distinction.)

Shagrir (2001; 2022) and Sprevak (2010) argue that functional indeterminacy can only be resolved by appeal to semantic content, but I suggest that the indeterminacy will often be resolved by appeal to non-semantic features of the explanatory context, in particular, by reference to the *use* to which a system is put for a task or cognitive capacity, say, to detect edges or to control motor behavior. Distinct realization functions will specify distinct causal organization supporting different counterfactuals about the behavior of the system in various circumstances, subserving the capacity in question. See Dewhurst (2018) for a similar idea. Both kinds of indeterminacy, then, will be resolved in favor of the mapping that does the best job of characterizing the causal processes responsible for the capacity to be explained, where "best job" will typically be determined by a variety of empirical and pragmatic considerations.

14. Indeed, as we shall see in chapter 3, that is the primary role played by content attribution in commonsense psychology as well.

15. See the papers in Piccinini (2012) for more on this topic.

16. See, for example, Chomsky (1994; 1995), though the distinction is recognized by many others. See Papineau (2015) for further discussion. Chomsky himself endorses only methodological naturalism, claiming that metaphysical (ontological) naturalism has had no coherent formulation since the demise of Descartes's corpuscular

mechanics in the early eighteenth century. As he puts it, "Newton eliminated the problem of 'the ghost in the machine' by eliminating the machine; the ghost was unaffected" (1994, 188).

17. To the extent that the *phenomenal intentionality research program* relies on introspection for the investigation of content, it violates methodological naturalism.

18. Clark (2015) promotes *predictive processing* as a general program for understanding the mind. My focus here is on his suggestion for naturalizing intentionality. Predictive processing is simply one proposal for how that might be accomplished.

19. Recall from the earlier discussion that the mathematical contents determined by element (1) are already naturalized.

20. Following convention, I use upper case to denote uninterpreted structures.

21. See also Bokulich (2003) and Morgan (2001).

Chapter 3

1. For explicit articulation and defense of the view, see Fodor (1975; 1979; 1987; 2008), Field (1978), and more recently Quilty-Dunn and Mandelbaum (2018), though many others endorse the view.

2. The RTM requires supplementation with an account of how MRs get their content. See, for example, the various proposals that fall under the "naturalistic semantics project," discussed in chapters 1 and 2.

3. See Fodor (1987) ("There is, so far as I know no human group that doesn't explain behavior by imputing beliefs and desires to the behavior. [And if an anthropologist claimed to have found such a group, I wouldn't believe him.]" p. 132); also Fodor (1992); Leslie (1987); Scholl and Leslie (1999).

4. The striking parallelism claim holds only if one sticks to mental states of a single type, in particular, to *beliefs*. Jones's believing that he has an important meeting in the morning may cause him to hope that he gets a good night's sleep tonight, but what is the inferential

relation between contents that "mirrors" the causal relation in this case?

5. Schwitzgebel (2013; 2021), making a similar point against a deep theory construal, argues that CP is theoretically *superficial*, though our arguments for the claim are different.

6. See Levin (2018) for a discussion of the distinction. *Realizer functionalism* is a "deep-state" view.

7. Hence, the dispositionalism at issue is *not* behaviorism.

8. In the same sense that a tree's rings carry information about its age.

9. McGeer (2021) identifies at least three types of constraints or norms:

1. *Evidential constraints* governing what it is appropriate to believe, given particular sorts of evidence.
2. *Evaluative constraints* governing what it is appropriate to desire, given our basic needs as humans, our individual circumstances, and so on.
3. *Executive constraints* governing what it is appropriate to prefer and do in light of one's beliefs and desires.

10. Though sometimes we simply *withdraw* an attitude attribution in the face of a failed prediction—maybe my friend Marisa didn't believe that we had an appointment after all; she must have forgotten—but if there is a pattern of nonconformity with the norm, we tend to criticize and withhold cooperation, friendship, and so on.

11. What socializes us to be good models of CP may be much broader than CP itself. We needn't assume that there is some precise way of demarcating the norms of CP from other social norms, for example, from rules of etiquette.

12. Evidence that belief/desire psychology (CP) is not universal, and hence at least in part a product of enculturation, includes but is not limited to: (1) linguistic evidence that some languages, notably *evidential languages*, generally lack propositional attitude verbs. Quechuan, for example, appears to have nothing close to a lexicalized concept for *belief*. Subjects use a verb meaning *say* to report the views of others. Jesuits missionaries introduced "credo" ("I believe") to teach Quechuan subjects to recite the Apostle's Creed, but there is little evidence of

its use outside religious contexts (See the discussion in Vinden 1996); (2) developmental evidence from children suggesting that the standard sequence of cognitive milestones—the acquisition of concepts for *wants* and *beliefs*, the development of *mind-reading* capacities—is scrambled in some cultures (e.g., Lillard 1997; 1998; Slaughter and Perez-Zapata 2014). In a striking study, Quechuan children tested up to age eight seemed to have no difficulty mastering the appearance/reality distinction at appropriate ages but could not master the false belief task; they appeared unable to attribute beliefs to others (see Vinden 1996). For summaries of the cross-cultural evidence and general discussion of the issue, see Mills (2001) and Strijbos and de Bruin (2013).

13. Modulo the worry raised in note 4.

14. The view that propositional attitude attributions function like measure predicates is developed and defended in Matthews (2007; 2011).

15. See Swoyer (1991) for more on this idea.

16. In computational cognitive science, the causal complexes (as I call them here) responsible for our cognitive capacities are specified by the "theory proper" (see chapter 1). Specifying the structures and processes underlying cognition is the job of scientific psychology, not of common sense.

17. There is trouble only when, as metaphysicians of belief, we come to think of beliefs as literally having intentional properties.

18. Recall the discussion of directedness introduced in chapter 2. As I noted there, directedness is not as fine-grained as the reference relation. For example, it doesn't cut between rabbits and temporal stages of rabbits.

19. See Van Leeuwen, Weisman, and Luhrmann (2021) for empirical support for this idea.

20. Recall from chapter 1 that a similar function is served by an intentional gloss of a computational theory.

21. Schwitzgebel (2013; 2021) calls implicit bias "in between" cases—that is, cases that are in between believing and not believing.

22. It might be thought that my causal complexes are similar to Dennett's (1991) "real patterns." But Dennett's patterns are most plausibly understood as patterns *in behavior* and cannot play a role in causing (and hence explaining) behavior. They are *abstracta*.

23. See the distinction cited earlier between *role functionalism* and *realizer (filler) functionalism*.

24. Thanks to Francois Recanati and Pierre Jacob for helpful discussion of the issues in this section.

25. The attribution of attitudes to nonhuman animals does not obviously serve a regulative function.

26. Davidson himself denied that animals have beliefs, because they don't have the *concept* of belief. In his view, having concepts, and hence having beliefs, requires having a language.

27. See the next section for another example.

28. Zimmerman (2018) also claims that our beliefs sometimes have indeterminate contents. He cites as an example Roman Catholic beliefs about the Trinity. "The indeterminacy of a belief's content is no argument against its psychological reality" (61).

29. The example, of course, is from Putnam (1973; 1975), though Putnam argues only that the vocable "water" in the mouths of Oscar and his twin has different meanings. On the assumption that a subject's utterances express his thoughts, the twins' thoughts have distinct contents, in virtue of the distinct concepts expressed by their use of the vocable "water."

Chapter 4

1. The view is sometimes called *intentionalism*. Proponents include Byrne (2001), Dretske (1995; 2003), Harman (1990), Lycan (1996; 2001), Tye (1995; 2000), and many others.

2. The terminology in this paragraph is from Byrne (2001), though Byrne calls representationalism "intentionalism."

3. Byrne (2001, 206) cites Tye (1995, 2000) as an utterly unrestricted representationalist, and says "Dretske, Harman, Lycan, and Shoemaker at least come close."

4. A view that claims simply that (all, some) experiences have content but does not claim that content determines phenomenal character is obviously weaker still, and is not generally considered a version of representationalism. This view is compatible with the position defended in this chapter.

5. Though see chapters 1 and 2 for argument that content attribution in cognitive science is motivated in part by pragmatic considerations and so is best described as a kind of *gloss*.

6. See Lycan (2015) for a dissenting view about whether adverbialism is a linguistic thesis.

7. External sortalism is similar in important respects to Jacobson and Putnam's (2014) *attributeism*.

8. See the discussion of inverted spectra in the next section for further relevant discussion of this point.

9. As Sellars (1956) noted, "looks" talk is often a way of hedging our descriptions of the external world.

10. See Harman (1990). As McLaughlin (2016) notes, the transparency point can be traced to Moore (1903) and is later noted by Broad (1925). Jacobson and Putnam (2014) point out that Reichenbach (1938) made a similar point, though he does not use the term "transparency."

11. For discussion of sounds and auditory experience, see O'Callaghan (2007).

12. For discussion of touch and haptic experience, see Fulkerson (2014).

13. For discussion of smell, see McGee (2020) and Barwich and Smith (2022). For discussion of both smell and taste, see Smith (2015).

14. For discussion of perspectival perception, see Green and Schellenberg (2017) and Lande (2018).

15. Recall that the objects and properties that constitute the PE framework (S2) are distinct from the objects and properties that constitute the EW framework (S1). It is *relational structure* that the two frameworks share.

16. In the second edition of the book (Fish 2021), Fish fails to mention sortal versions of adverbialism at all, focusing exclusively on qualia adverbialism and the phenomenal intentionality program, both of which attempt to ground experiential properties in phenomenal properties. This idea is antithetical to the spirit of the original adverbialism proposal advocated by Sellars and Cornman.

17. See Smart's (1959) account of his "topic-neutral" analyses.

18. There is much to be said about the role of language in regularizing experience, but the topic is beyond the scope of this chapter.

19. D'Ambrosio (2019) proposes a linguistic solution to the many properties problem. I will not discuss his solution here. External sortalism's advantage in avoiding the many properties problem depends on the view not being a linguistic thesis, and so does not require a linguistic solution.

20. See Morrison (2020) for defense of this idea.

References

Allen, C. (2013). "The Geography of Partial Understanding." *American Philosophical Quarterly* 50, No. 3: 249–262.

Barrett, L. (2011). *Beyond the Brain*. Princeton University Press.

Barwich, A-S., & Smith, B. C. (2022). "From Molecules to Perception: Philosophical Investigations of Smell." *Philosophy Compass* 17, No. 11: e12883.

Bennett, M. R., & Hacker, P. M. S. (2003). *Philosophical Foundations of Neuroscience*. Wiley-Blackwell.

Berkeley, G. (1709/2009). "An Essay Towards a New Theory of Vision." In D. Clarke (Ed.), *Berkeley: Philosophical Writings* (pp. 1–65). Cambridge University Press.

Bokulich, A. (2003). "Horizontal Models: From Bakers to Cats." *Philosophy of Science* 70:609–627.

Broad, C. D. (1923). *Scientific Thought*. Harcourt, Brace and Company.

Broad, C. D. (1925). *The Mind and Its Place in Nature*. Routledge and Kegan Paul.

Buchanan, R., & Grzankowski, A. (2022). "Why Bother with Mental Representations: Review of *What Are Mental Representations?*" *Metascience* 31, No. 3: 415–418.

Burge, T. (2010). *Origins of Objectivity*. Oxford University Press.

Byrne, A. (2001). "Intentionalism Defended." *The Philosophical Review* 110:199–240.

Cash, M. (2008). "The Normativity Problem: Evolution and Naturalized Semantics." *The Journal of Mind and Behavior* 29:99–138.

Chirimuuta, M. (2021). "Your Brain Is Like a Computer: Function, Analogy, Simplification." In F. Calzavarini & M. Viola (Eds.), *Neural Mechanisms: New Challenges in the Philosophy of Neuroscience* (Studies in Brain and Mind 17, pp. 235–261). Springer.

Chirimuuta, M. (2024). *The Brain Abstracted: Simplification in the History and Philosophy of Neuroscience*. MIT Press.

Chomsky, N. (1994). "Naturalism and Dualism in the Study of Language and Mind." *International Journal of Philosophical Studies* 2:181–209.

Chomsky, N. (1995). "Language and Nature." *Mind* 104:1–61.

Clark, A. (2015). "Predicting Peace: The End of the Representation Wars." In T. Metzinger & J. Windt (Eds.), *Open MIND*. MIT Press. http://doi.org/10.25358/openscience-454

Colombo, M., & Series, P. (2012). "Bayes in the Brain: On Bayesian Modelling in Neuroscience." *British Journal for the Philosophy of Neuroscience* 63, No. 3: 697–723.

Cornman, J. W. (1971). *Materialism and Sensations*. Yale University Press.

Crane, T. (1992). "The Nonconceptual Content of Experience." In T. Crane (Ed.), *The Contents of Experience: Essays on Perception* (pp. 136–157). Cambridge University Press.

Crane, T. (2000). "The Origins of Qualia." In T. Crane and S. Patterson (Eds.), *The History of the Mind-Body Problem* (pp. 169–194). Routledge.

Crane, T., & French, C. (2021). "The Problem of Perception." *Stanford Encyclopedia of Philosophy*. E. Zalta (Ed.). https://plato.stanford.edu/archives/fall2021/entries/perception-problem/

Craver, C. (2006). "When Mechanistic Models Explain." *Synthese* 153:355–376.

Cummins, R. (1989) *Meaning and Mental Representation.* MIT Press.

D'Ambrosio, J. (2019). "A Pure Perceptual Adverbialism." *The Journal of Philosophy* 116:414–446.

Davidson, D. (1982). "Rational Animals." *Dialectica* 36, No. 2: 317–327.

Dewhurst, J. (2018). "Individuation without Representation." *The British Journal for the Philosophy of Science* 69, No. 1: 103–116.

Dennett, D. (1978). *Brainstorms.* MIT Press.

Dennett, D. (1991). "Real Patterns." *The Journal of Philosophy* 88:27–51.

Dinges, A. (2015). "The Many-Relations Problem for Adverbialism." *Analysis* 75, No. 2: 231–237.

Dretske, F. (1981). *Knowledge and the Flow of Information.* MIT Press.

Dretske, F. (1995). *Naturalizing the Mind.* MIT Press.

Dretske, F. (2003). "Experience as Representation." In E. Villanueva (Ed.), *Philosophical Issues 13: Philosophy of Mind* (pp. 67–82). Ridgeview.

Ducasse, C. J. (1942). "Moore's Refutation of Idealism." In P. Schlipp (Ed.), *The Philosophy of G.E. Moore* (Vol. I, pp. 225–251). Open Court.

Egan, F. (2017). "Function Theoretic Explanation and the Search for Neural Mechanisms." In D. M. Kaplan (Ed.), *Explanation and Integration in Mind and Brain Science* (pp. 145–163). Oxford University Press.

Egan, F. (2020). "Content Is Pragmatic: Comments on Nicholas Shea's *Representation in Cognitive Science.*" *Mind and Language* 35:368–376.

Egan, F. (2022). "The Elusive Role of Normal-Proper Function in Cognitive Science." *Philosophy and Phenomenological Research* 105:468–475.

Facchin, M. (2022). "Troubles with Mathematical Contents." *Philosophical Psychology*, 1–24. https://doi.org/10.1080/09515089.2022.2119952

Facchin, M. (2023). "Neural Representations Unobserved—Or: A Dilemma for the Cognitive Neuroscience Revolution." *Synthese* 203, No. 7. https://doi.org/10.1007/s11229-023-04418-6

Field, H. (1978). "Mental Representation." *Erkenntnis* 13:9–61. Reprinted in N. Block (Ed.), *Readings in the Philosophy of Psychology* (pp. 78–114). Harvard University Press, 1981.

Fish, W. (2010). *Philosophy of Perception: A Contemporary Introduction.* Routledge.

Fish, W. (2021). *Philosophy of Perception: A Contemporary Introduction* (2nd ed.). Routledge.

Flavell, J. (1986). "The Development of Children's Knowledge about the Appearance–Reality Distinction." *American Psychologist* 41, No. 4: 418–425.

Flavell, J., Green, F., & Flavell, E. (1986). "Development of Knowledge about the Appearance-Reality Distinction." *Monographs of the Society for Research in Child Development* 51, No. 1: 1–87.

Fodor, J. A. (1975). *The Language of Thought.* Thomas Y. Crowell Co.

Fodor, J. A. (1979). "Propositional Attitudes." In *RePresentations: Essays on the Foundations of Cognitive Science.* MIT Press.

Fodor, J. A. (1987). *Psychosemantics: The Problem of Meaning in the Philosophy of Mind.* MIT Press.

Fodor, J. A. (1990). *A Theory of Content and Other Essays.* MIT Press.

Fodor, J. A. (1992). "A Theory of the Child's Theory of Mind." *Cognition* 44:283–296.

Fodor, J. A. (2000). *The Mind Doesn't Work That Way.* MIT Press.

Fodor, J. A. (2008). *LOT 2: The Language of Thought Revisited.* Oxford University Press.

Fulkerson, M. (2014). *The First Sense: A Philosophical Study of Human Touch.* MIT Press.

Gallistel, C. R. (1990). *The Organization of Learning.* MIT Press.

Garson, J. (2019). *What Biological Functions Are and Why They Matter.* Cambridge University Press.

Gergen, K. (1985). "The Social Constructivist Movement in Modern Psychology." *American Psychologist* 40:266–275.

Gergen, K. (1997). "The Place of the Psyche in a Constructed World." *Theory and Psychology* 7:723–746.

Green, E. J., & Schellenberg, S. (2017). "Spatial Perception: The Perspectival Aspect of Perception." *Philosophy Compass* 13, No. 2: e12472.

Grzankowski, A. (2017). "The Determinable-Determinate Relation Can't Save Adverbialism." *Analysis* 78, No. 1: 45–52.

Harman, G. (1990). "The Intrinsic Character of Experience." In J. E. Tomberlin (Ed.), *Philosophical Perspectives 4: Action Theory and the Philosophy of Mind* (pp. 31–52). Ridgeview.

Hartmann, S. (1999). "Models and Stories in Hadron Physics." In M. Morgan & M. Morrison (Eds.), *Models as Mediators. Perspectives on Natural and Social Science* (pp. 326–346). Cambridge University Press.

Horgan, T., & Graham, G. (2012). "Phenomenal Intentionality and Content Determinacy." In R. Schantz (Ed.), *Prospects for Meaning* (pp. 321–344). De Gruyter.

Hutto, D., & Myin, E. (2012). *Radicalizing Enactivism: Basic Minds without Content*. MIT Press.

Hutto, D., and Myin, E. (2017). *Evolving Enactivism: Basic Minds Meet Content*. MIT Press.

Hutto, D., and Myin, E. (2020). "Deflating Deflationism about Mental Representation." In J. Smortchkova, K. Dolega, & T. Schlicht (Eds.), *What Are Mental Representations?* (pp. 79–100). Oxford University Press.

Ismael, J. (2013). "Naturalism on the Sydney Plan." In M. Haug (Ed.), *Philosophical Methodology: The Armchair or the Laboratory?* (pp. 86–104). Routledge.

Jackson, F. (1975). "On the Adverbial Analysis of Visual Experience." *Metaphilosophy* 6, No. 2: 127–135.

Jacob, P. (2023). "Intentionality." *Stanford Encyclopedia of Philosophy*. E. Zalta & U. Nodelman (Eds.). https://plato.stanford.edu/archives/spr2023/entries/intentionality/

Jacobson, H., & Putnam, H. (2014). "The Needlessness of Adverbialism, Attributeism and Its Compatibility with Cognitive Science." *Philosophia* 42:555–570.

Klein, C. (2015). *What the Body Commands: The Imperative Theory of Pain*. MIT Press.

Kriegel, U. (2008). "The Dispensability of (Merely) Intentional Objects." *Philosophical Studies* 141:79–95.

Kriegel, U. (2013). "The Phenomenal Intentionality Research Program." In *Phenomenal Intentionality* (pp. 1–26). Oxford University Press.

Lande, K. (2018). "The Perspectival Character of Perception." *The Journal of Philosophy* 115:187–214.

Leslie, A. (1987). "Pretense and Representation: The Origins of 'Theory of Mind.'" *Psychological Review* 94:412–426.

Lettvin, J., Maturana, H., McCulloch, W., & Pitts, W. (1959). "What the Frog's Eye Tells the Frog's Brain." *Proceedings of the IRE* 47, No. 11: 1940–1951.

Levin, J. (2018). "Functionalism." *Stanford Encyclopedia of Philosophy*. E. Zalta & U. Nodelman (Eds.). https://plato.stanford.edu/archives/sum2023/entries/functionalism/

Levine, J. (1983). "Materialism and Qualia: The Explanatory Gap." *Pacific Philosophical Quarterly* 64:354–361.

Lillard, A. (1997). "Other Folks' Theories of Mind and Behavior." *Psychological Science* 8:268–274.

Lillard, A. (1998). "Ethnopsychologies: Cultural Variations in Theories of Mind." *Psychological Bulletin* 123:3–32.

Luo, X., Mok, R., & Love, B. (2024). "The Inevitability and Superfluousness of Cell Types in Spatial Cognition." *BioRxiv* 2024.01.10. https://www.biorxiv.org/content/10.1101/2024.01.10.575026v1

Lycan, W. (1996). *Consciousness and Experience*. MIT Press.

Lycan, W. (2001). "The Case for Phenomenal Externalism." In J. E. Tomberlin (Ed.), *Metaphysics (Philosophical Perspectives*, Vol. 15). Ridgeview Publishing.

Lycan, W. (2015). "Representational Theories of Consciousness." *Stanford Encyclopedia of Philosophy*, E. Zalta & U. Nodelman (Eds.). https://plato.stanford.edu/archives/win2023/entries/consciousness-representational/

Marr, D. (1982). *Vision*. Freeman.

Malcolm, N. (1973). "Thoughtless Brutes." *Proceedings and Addresses of the American Philosophical Association* 46:5–20.

Marcus, M. (1980). *A Theory of Syntactic Recognition for Natural Language*. MIT Press.

Matthen, M. (1988). "Biological Functions and Perceptual Content." *Journal of Philosophy* 85:5–27.

Matthews, R. (2007). *The Measure of Mind: Propositional Attitudes and Their Attribution*. Oxford University Press.

Matthews, R. (2011). "Measurement-Theoretic Accounts of Propositional Attitudes." *Philosophy Compass* 6, No. 11: 828–841.

Matthews, R. (2013). "Belief and Belief's Penumbra." In N. Nottlemann (Ed.), *New Essays on Belief* (pp. 100–123). Palgrave Macmillan.

McGee, H. (2020). *Nose Dive: A Field Guide to the World's Smells*. Penguin Press.

McGeer, V. (2007). "The Regulative Dimension of Folk Psychology." In D. Hutto & M. Radcliffe (Eds.), *Folk Psychology Reassessed* (pp. 137–156). Springer.

McGeer, V. (2021). "Enculturating Folk Psychologists." *Synthese* 199:1039–1063.

McLaughlin, B. (2016). "Normal Geometrical Misperception: The Skewed View from Here." *Philosophical Topics* 44:231–299.

Mendelovici, A. (2014). "Pure Intentionalism about Moods and Emotions." In U. Kriegel (Ed.), *Current Controversies in Philosophy of Mind* (pp. 135–157). Routledge.

Mendelovici, A. (2018). *The Phenomenal Basis of Intentionality*. Oxford University Press.

Mendelovici, A., & Bourget, D. (2014). "Naturalizing Intentionality: Tracking Theories versus Phenomenal Intentionality Theories." *Philosophy Compass* 9, No. 5: 325–337.

Millikan, R. (1984). *Language, Thought, and Other Biological Categories*. MIT Press.

Millikan, R. (1989). "Biosemantics." *The Journal of Philosophy* 86: 281–297.

Millikan, R. (1991). "Speaking Up for Darwin." In B. Loewer & G. Rey (Eds.), *Meaning in Mind: Fodor and his Critics* (pp. 151–164). Blackwell.

Mills, S. (2001). "The Idea of Different Folk Psychologies." *International Journal of Philosophical Studies* 4:501–519.

Moore, G. E. (1903). "The Refutation of Idealism." *Mind* 12:433–453.

Morgan, A. (2014). "Representations Gone Mental." *Synthese* 191, No. 2: 213–244.

Morgan, M. (2001). "Models, Stories and the Economic World." *Journal of Economic Methodology* 8, No. 3: 361–384.

Morrison, J. (2020). "Perceptual Variation and Structuralism." *Nous* 54:290–326.

Neander, K. (1995). "Misrepresenting and Malfunctioning." *Philosophical Studies* 79:109–141.

Neander, K. (2017). *A Mark of the Mental: In Defense of Informational Teleosemantics*. MIT Press.

Newen, A., & Vosgerau, G. (2020). "Situated Mental Representations: Why We Need Mental Representations and How We Should

Understand Them." In J. Smortchkova, K. Dolega, & T. Schlicht (Eds.), *What Are Mental Representations?* (pp. 178–212). Oxford University Press.

O'Callaghan, C. (2007). *Sounds*. Oxford University Press.

O'Brien, G., & Opie, J. (2004). "Notes Toward a Structuralist Theory of Mental Representation." In H. Clapin, P. Staines, & P. Slezak (Eds.), *Representation in Mind: New Approaches to Mental Representation* (pp. 1–20). Elsevier.

O'Keefe, J. A. (1976). "Place Units in the Hippocampus." *Experimental Neurology* 51:78–109.

O'Keefe, J. A., & Dostrovsky, J. (1971). "The Hippocampus as a Spatial Map: Preliminary Evidence from Unit Activity in the Freely Moving Rat." *Brain Research* 34:171–175.

O'Keefe, J. A., & Nadel, L. (1978). *The Hippocampus as a Cognitive Map*. Oxford University Press.

Papayannopoulos, P., Fresco, N., & Shagrir, O. (2022). "On Two Different Kinds of Computational Indeterminacy." *The Monist* 105:229–246.

Papineau, D. (1987). *Reality and Representation*. Basil Blackwell.

Papineau, D. (1993). *Philosophical Naturalism*. Basil Blackwell.

Papineau, D. (2015). "Naturalism." *The Stanford Encyclopedia of Philosophy*. E. Zalta & U. Nodelman (Eds.). https://plato.stanford.edu/archives/fall2023/entries/naturalism/

Piccinini, G., & Craver, C. (2011). "Integrating Psychology and Neuroscience: Functional Analyses as Mechanism Sketches." *Synthese* 183, No. 3: 283–311.

Piccinini, G., ed. (2012). "Special Issue: A Computational Foundation for the Study of Cognition II." *Journal of Cognitive Science* 13, No. 1.

Putnam, H. (1973). "Meaning and Reference." *Journal of Philosophy* 70:699–711.

Putnam, H. (1975). "The Meaning of Meaning." in *Mind, Language, and Reality: Philosophical Papers II* (pp. 215–271). Cambridge University Press.

Putnam, H. (1988). *Representation and Reality*. MIT Press.

Quilty-Dunn, J., & Mandelbaum, E. (2018). "Against Dispositionalism: Belief in Cognitive Science." *Philosophical Studies* 175:2353–2372.

Quine, W. V. O. (1960). *Word and Object*. MIT Press.

Ramsey, W. (2007). *Representation Reconsidered*. Cambridge University Press.

Ramsey, W. (2020). "Defending Representational Realism." In J. Smotchkova, K. Dolega, & T. Schlicht (Eds.), *What Are Mental Representations?* (pp. 54–78). Oxford University Press).

Ranek, J. B. (1973). "Studies on Single Neurons in Dorsal Rats. Part I. Behavioral Correlates and Firing Repertoires." *Experimental Neurology* 41:462–531.

Reichenbach, H. (1938). *Experience and Prediction. An Analysis of the Foundations and the Structure of Knowledge*. University of Chicago Press.

Robinson, H. (1994). *Perception*. Routledge.

Ryder, D. (2004). "SINBaD Neurosemantics: A Theory of Mental Representation." *Mind and Language* 19:211–240.

Ryle, G. (1949). *The Concept of Mind*. University of Chicago Press.

Scholl, B., & Leslie, A. (1999). "Modularity, Development, and 'Theory of Mind.'" *Mind and Language* 14:131–153.

Schwitzgebel, E. (2013). "A Dispositional Approach to the Attitudes: Thinking outside the Belief Box." In N. Nottlemann (Ed.), *New Essays on Belief* (pp. 75–99). Palgrave Macmillan.

Schwitzgebel, E. (2021). "The Pragmatic Metaphysics of Belief." In C. Borgoni, D. Kindermann & A. Onofri (Eds.), *The Fragmented Mind* (pp. 350–376). Oxford University Press.

Searle, J. (1991). *The Rediscovery of the Mind*. MIT Press.

Sellars, W. (1956). "Empiricism and the Philosophy of Mind." In H. Feigl (Ed.), *Minnesota Studies in the Philosophy of Science* (Vol. I, pp. 307–329). University of Minnesota Press.

Sellars, W. (1963). "Philosophy and the Scientific Image of Man." In *Science, Perception and Reality* (pp. 1–40). Routledge and Kegan Paul.

Sellars, W. (1975). "The Adverbial Theory of the Objects of Sensation." *Metaphilosophy* 6, No. 2: 144–160.

Seung, S. (1996). "How the Brain Keeps the Eyes Still." *Proceedings of the National Academy of Science USA* 93:13339–13344.

Seung, S. (1998). "Continuous Attractors and Oculomotor Control." *Neural Networks* 11:1253–1258.

Seung, S., Lee, D., Reis, B., & Tank, D. (2000). "Stability of the Memory of Eye Position in a Recurrent Network of Conductance based Model Neurons." *Neuron* 26:259–271.

Shadmehr, R., & Wise, S. (2005). *The Computational Neurobiology of Reaching and Pointing: A Foundation for Motor Learning*. MIT Press.

Shagrir, O. (2001). "Content, Computation and Externalism." *Mind* 110:369–400.

Shagrir, O. (2012). "Structural Representations and the Brain." *British Journal for the Philosophy of Science* 63, No. 3: 519–545.

Shagrir, O. (2022). *The Nature of Physical Computation*. Oxford University Press.

Shea, N. (2013). "Naturalising Representational Content." *Philosophy Compass* 8, No. 5: 496–509.

Shea, N. (2018). *Representation in Cognitive Science*. Oxford University Press.

Siegel, S. (2016). "The Contents of Perception." *Stanford Encyclopedia of Philosophy*. E. Zalta (Ed.). https://plato.stanford.edu/archives/fall2021/entries/perception-contents/

Slaughter, V. & Perez-Zapata, D. (2014). "Cultural Variations in the Development of Mind Reading." *Child Development Perspectives* 8, No. 4: 237–241.

Slezak, P. (2023). *Spectator in the Cartesian Theater: Where Theories of Mind Went Wrong Since Descartes*. Rowman & Littlefield.

Smart, J. J. C. (1959). "Sensations and Brain Processes." *The Philosophical Review* 68, No. 2: 141–156.

Smith, Barry C. (2015). "The Chemical Senses." In M. Matthen (Ed.), *The Oxford Handbook of Philosophy of Perception* (pp. 314–352). Oxford University Press.

Smith, Brian C. (2002). "The Foundations of Computing." In M. Scheutz (Ed.), *Computationalism: New Directions* (pp. 23–58). MIT Press.

Sprevak, M. (2010). "Computation, Individuation, and the Received View on Representation." *Studies in History and Philosophy of Science Part A* 41, No. 3: 260–270.

Sprevak, M. (2013). "Fictionalism about Neural Representations." *The Monist* 96:539–560.

Stalnaker, R. (2008). *Our Knowledge of the Internal World*. Oxford University Press.

Sterelny, K. (1990). *The Representational Theory of Mind: An Introduction*. Blackwell.

Stich, S. (1983). *From Folk Psychology to Cognitive Science: The Case against Belief*. MIT Press.

Strijbos, D., & de Bruin, L. (2013). "Universal Belief-Desire Psychology: A Dilemma for Theory Theory and Simulation Theory." *Philosophical Psychology* 26:744–764.

Swoyer, C. (1991). "Structural Representation and Surrogative Reasoning." *Synthese* 87:449–508.

Toon, A. (2022). "Fictionalism and Intentionality." In T. Demeter, T. Parent, & A. Toon (Eds.), *Mental Fictionalism: Philosophical Explorations* (pp. 52–69). Routledge.

Toon, A. (2023). *Mind as Metaphor: A Defence of Mental Fictionalism* Oxford University Press.

Tye, M. (1975). "The Adverbial Theory: A Defense of Sellars against Jackson." *Metaphilosophy* 6, No. 2: 136–143.

Tye, M. (1984). "The Adverbial Approach of Visual Experience." *The Philosophical Review* 93:195–226.

Tye, M. (1995). *Ten Problems of Consciousness*. MIT Press.

Tye, M. (2000). *Consciousness, Color, and Content*. MIT Press.

Ullman, S. (1979). *The Interpretation of Visual Motion*. MIT Press.

Van Leeuwen, N., Weisman, K., & Luhrmann, T. M. (2021). "To Believe Is Not to Think: A Cross-Cultural Finding." *Open Mind: Discoveries in Cognitive Science* 5:91–99.

Vinden, P. (1996). "Junin Quechua Children's Understanding of Mind." *Child Development* 67:1707–1716.

Vygotsky, L. (1978). *Mind in Society: The Development of Higher Psychological Processes*. Harvard University Press.

Zawidzki, T. (2008). "The Function of Folk Psychology: Mind Reading or Mind Shaping?" *Philosophical Explorations* 11, No. 3: 193–210.

Zednik, C., & Jakel, F. (2016). "Bayesian Reverse-Engineering Considered as a Research Strategy for Cognitive Science." *Synthese* 193, No. 12: 3951–3985.

Zimmerman, A. (2018). *Belief: A Pragmatic Picture*. Oxford University Press.

Index

Publisher contact:
The MIT Press
Massachusetts Institute of Technology
77 Massachusetts Avenue, Cambridge, MA 02139
mitpress.mit.edu

EU Authorised Representative:
Easy Access System Europe, Mustamäe tee 50,
10621 Tallinn, Estonia
gpsr.requests@easproject.com

Printed by Integrated Books International,
United States of America